Quan
The Power of the Mind

Discover all the important features of Quantum Physics and the Law of Attraction. Find out how it really works to change your life for the better.

Nancy Patterson

©Copyright 2020 Nancy Patterson All rights reserved

The content within this book may not be reproduced, duplicated, or transmitted without direct written permission from the author or the publisher.

Under no circumstances will any blame or legal responsibility be held against the publisher, or author, for any damages, reparation, or monetary loss due to the information contained within this book, either directly or indirectly.

Legal Notice

This book is copyright protected. This book is only for personal use. You cannot amend, distribute, sell, use, quote, or paraphrase any part, or the content within this book, without the consent of the author-publisher.

Disclaimer Notice

Please note that the information contained within this document is for educational and entertainment purposes only. All effort has been executed to present accurate, up-to-date, and reliable, complete information. No warranties of any kind are declared or implied. Readers acknowledge that the author is not engaging in the rendering of legal, financial, medical, or professional advice.

TABLE OF CONTENTS

Introduction 4
Max Planck, the Father of Quantum Theory 9
Quantum Origins of the Universe 17
Classical Physics vs. Quantum Physics 27
Rutherford's Experiment 43
Quantum Tunnelling 52
The Black Body Radiation 54
Light 63
Quantum Physics and the Law of Attraction 69
Photoelectric Effect: Einstein's Theory 77
The Theory of Relativity 86
Quantum Physics and Waves 93
The Heisenberg's Uncertainty Law 99
Quantum Super-Positioning 101
Quantum Computing 103
How Quantum Physics Affects You 110
Conclusion 124

Introduction

Quantum mechanics, commonly called quantum physics, is the relationship between energy and matter. The word 'quantum' is Latin for 'how much.' Mechanics refers to a unit in which quantum theory assigns a measure to specific physical quantities in minimal quantities. In essence, quantum expressions are generally visualized and studied sub atomically with subatomic particles. Subatomic particles are small. If an atom were the size of a house, the subatomic molecule would be the size of a wad of chewing gum in the kitchen cabinet of that house. There were a few things that needed to occur before the investigation of quantum mechanics flourished. In 1838 cathode beams were disclosed and in 1850 Gustav Kirchhoff published a statement on the problem of 'blackbody radiation.' Soon after, in 1877, Ludwig Boltzmann proposed that the energetic states of a physical system could be unconnected.

In 1900, Max Planck developed the theory that energy is radiated and absorbed. He made an equation known as 'Planck's Activity Consistent.' Planck is known as the grandfather of quantum materials science. After his theory was circulated, other scientists took note of it and discovered other theoretical structures, until eventually quantum mechanics was theorized and studied around the world. Thanks to quantum physics, we discovered gravity, we have superconductors and magnetic resonance imaging equipment in hospitals, and now we can even see that time travel is possible.

It all sounds so fantastic, but scientists in the field of quantum mechanics will tell you. It is hard for most of us to understand the connection between subatomic particles and the law of attraction. During the research of quantum mechanics, it was discovered that subatomic particles determine the direction in which the Earth rotates. Another force moves these particles of physical matter out into the universe. After some double-blind slit tests using subatomic particles as subjects, it was discovered that they could switch between wave-shaped particles and then back to block-shaped particles again. These particles could leave our dimension and reenter. We also found that these subatomic particles deliberately changed from wave-shaped particles according to purpose. We found out that when we were testing the particles, we could not remove ourselves from the equation. We influenced the particles by thinking about the result. This is where it becomes confusing, the concept confused Einstein until his death. The duality of particles and waves is not easy for most of us to comprehend.

One of the theories that emerged from the foundations of quantum physics is that we manipulate the fabric of life by thinking about it. Our thoughts have an expression that comes out and therefore brings us to what we focus on to make it a reality. This is the law of attraction. Every single quantum physicist will agree on one thing. The subatomic particles, energy packets or quantum, are not particles at a certain point in space and time such as a table or a chair, but they are just a probability that they could exist at different points in space and time.

The act of our observation turns it into a 'physical' particle at a certain point in space and time. Once we withdraw our attention from it, it becomes a probability again. Imagine that the sofa in your living room is a sizeable subatomic particle. This is how it would behave: If you are not at home and do not think about your sofa, it would 'disappear' and it would become a probability that it could reappear anywhere in your living room or anywhere else in the universe. If you came home thinking about sitting on the sofa in a specific place in your living room and looking for the sofa where you would like to relax, it would reappear! This seems like a fantasy, but it is a scientific fact that subatomic particles behave this way.

The presence of your sofa is only the result of your seeing it, expecting it, and deciding that it is there. It is not a completely independent existence. It doesn't have an entirely separate existence, regardless of the observer. The astonishing thing is that all matter is made up of large amounts of these particles. Therefore, all matter behaves precisely as a large group of subatomic particles would.

Quantum physics confirms that a thing can only exist if it is observed. The 'quanta' are organized according to the influence of the mind of the observers. When something is observed, the quanta merge into subatomic particles and then into atoms, followed by molecules, until finally something in the physical world manifests itself as a localized temporal space-time experience that can be perceived through our five physical senses. This leads to something that appears to be reliable and part of what people usually understand as physical reality.

An experiment by modern quantum physicists shows and proves that the entire universe exists through experience. The most fantastic research in quantum physics in recent scientific discovery is probably the double-slit experiment. Every single thought, as energy, directly and instantly influences the quantum field. Thus 'quanta' merge into a localized, observable experience event, an object, or other influence. This process is the basis for how everyone creates their own reality.

Those who understand and comply with universal laws are conscious creators, while others create their life experience unconsciously. As a result, they attribute everything they have experienced as a consequence of their unconscious thinking to superstitious beliefs such as luck, fate, chance or fortune. We know, however, that conscious creation is also the basis of the law of attraction and the law of cause and effect.

Max Planck, the Father of Quantum Theory

All objects emit electromagnetic radiation, which is called heat radiation. But we only see it when the objects are very hot, because they also emit visible light. Glowing iron or our sun would be examples of this. Physicists sought a formula that would correctly describe the emission of electromagnetic radiation but struggled to find it. At last, in 1900, the German physicist Max Planck (1858-1947) took a courageous step.

The emission of electromagnetic radiation means that there is emission of energy. According to Maxwell's equations, this release of energy should take place continuously, meaning that any value is possible for the energy output. Max Planck assumed that the energy output could only take place in multiple energy packets, i.e., in steps, which led him to the correct formula. Planck called these energy packets 'quanta.' Therefore, quantum theory was born in 1900.

It is important to note that only the emission (and the absorption) of the electromagnetic radiation should occur in the form of quanta. Planck, however, did not assume that it was composed of quanta, because that would have meant it would have a particle character. However, like all physicists of his time, he was convinced that electromagnetic radiation consisted exclusively of waves. Young's double-slit experiment revealed it, and Maxwell equations established it.

In 1905, an interloper named Albert Einstein was much bolder. He took a glance at the photoelectric outcome, or the way electrons could be knocked out of metals by irradiation with light. According to classical physics, the energy of the knocked-out electrons would depend on the intensity of the light. Strangely enough, this was not the case. The energy of the electrons did not depend on the intensity but on the frequency of the light. Einstein could explain that. For this, let's go back to Max Planck's quanta. The energy of each quantum depends on the frequency of the electromagnetic radiation. The higher the frequency, the greater the energy of the quantum. Einstein now assumed, in contrast to Planck, that the electromagnetic radiation itself consisted of quanta. The interaction of a single quantum with a single electron on the metal surface causes this electron to be knocked out. The quantum releases its energy to the electron. Therefore, the energy of the knocked-out electron depends on the frequency of the incident light.

The skepticism was great at first. This meant that electromagnetic radiation would have both a wave and a particle character. Another experiment showed its particle character. This experiment was conducted with X-rays and electrons by the American physicist Arthur Compton (1892-1962) in 1923. As already mentioned, X-rays are also electromagnetic radiations, but they have a much higher frequency than visible light.

Therefore, quanta in X-rays are very energetic, which is why they can colonize this particular form. That also makes them so threatening. Compton was adept at identifying those X-rays and electrons acted as billiard balls when they met. This again showed the bit character of the electromagnetic radiation. So, their twofold nature, the alleged "wave-particle dualism," was at last acknowledged. Compton introduced the term "photons" for the quanta of electromagnetic radiation.

What are photons? That is still unclear today. Under no circumstances should they be imagined as small spheres moving forward at the speed of light. The photons are not located in space; they are never at a certain place. In 1951, Albert Einstein said, "Fifty years of hard thinking have not brought me any closer to the answer to the question what are light quanta? Today, every Tom, Dick, and Harry imagines they know. But they're wrong." Today, this is still true.

The Bohr Atomic Model
We take atoms for granted. Their existence was still controversial until the early 20th century, but as early as the 5th century BCE, the ancient Greeks, especially Leucippus and his pupil Democritus, spoke of atoms. They thought that matter was made up of tiny, indivisible units. They called these atoms because in ancient Greek "Atomos" meant indivisible. In his miracle year 1905, Albert Einstein not only presented the special theory of relativity and solved the mystery of the photoelectric effect, but was also able to explain the

Brownian motion. In 1827, the Scottish botanist and physician Robert Brown (1773-1858) discovered that dust particles that were only visible under the microscope made jerky movements in water. Einstein was able to explain this by the fact that much smaller particles, which are not visible even under the microscope, collide in huge numbers with the dust particles and that they are subject to random fluctuations. The latter leads to jerky movements. The invisible particles must be molecules. Therefore, the explanation of the Brownian movement was considered as its validation and thus also as the validation of the atoms.

In 1897, the British physicist Joseph John Thomson (1856-1940) introduced the idea of electrons as a component of atoms and developed the first atomic model, known as the raisin cake model. According to this model, atoms consisted of an evenly distributed and positively charged mass in which the negatively charged electrons were embedded like raisins in a cake batter. This was refuted in 1910 by the New Zealand physicist Ernest Rutherford (1871-1937). With his experiments at the University of Manchester, he was able to show that atoms are almost empty. They were made up of a small, positively charged nucleus surrounded by electrons. They would rotate close to the nucleus as planets rotate around the sun. At the time, another form of movement was inconceivable. This led physics into a deep crisis. Since electrons had an electrical charge, their circular motion would cause them to release energy in the form of electromagnetic radiation. Therefore, the electrons should eventually fall into the nucleus. This prompted the deep crisis, because based on this, there would be no atoms at all.

In 1913, a young colleague of Ernest Rutherford, the Danish physicist Niels Bohr (1885-1962), tried to explain atoms' stability. He transferred the idea of quanta to the orbits of electrons in atoms. This meant that there were no random orbits around the nucleus for the electrons, but that only certain orbits were allowed. Each one had a certain energy. Bohr assumed that these allowed orbits to be stable because the electrons on them did not emit electromagnetic radiations. However, he was not able to explain why this was the case.

Nevertheless, his atomic model was initially quite successful because it could explain the Balmer formula. For some time, it had been identified that atoms only absorb light at certain frequencies. These are called spectral lines. In 1885, the Swiss mathematician and physicist Johann Jakob Balmer (1825-1898) found a formula that could correctly describe the frequencies of spectral lines. However, he could not explain them. Bohr then succeeded with his atomic model, at least for the hydrogen atom. This is because photons can be excited by photons, causing them to jump into higher energy orbits. This is the famous quantum leap, the smallest possible leap ever. Since only certain orbits are allowed in Bohr's atomic model, the energy and the frequency of the exciting photons must correspond exactly to the energy difference between the initial and the excited orbit. This explained Balmer's formula. Bohr's atomic model quickly reached its limits because it only worked for the hydrogen atom. The German physicist Arnold Sommerfeld (1868-1951) expanded it, but it still represented a rather unconvincing mix of classical physics with quantum aspects. t also still could not explain why certain orbits of the electrons would be stable. Sommerfeld had a young assistant, Werner Heisenberg (1901-1976), who, in his doctoral thesis, addressed the Bohr atom model extended by Sommerfeld. He wanted to improve it. In 1924, Heisenberg became assistant to Max Born (1882-1970) in Göttingen. The breakthrough occurred in 1925, on the island of Helgoland where he was being cured of hay fever. He explained the frequencies of the spectral lines, including their intensities, using so-called matrices. He published his theory in 1925 and his boss Max Born and Pascual Jordan (1902-1980) both approved it. This is considered to be the first quantum theory and it is called matrix mechanics. I will not explain it in more detail because it is not very clear and because there is an alternative mathematically equivalent to it. It is much more popular because it is easier to handle. It is called wave mechanics and was developed in 1926, just one year after matrix mechanics, by the Austrian physicist Erwin Schrödinger (1887-1961).

The Schrödinger Equation
Before we come to the wave mechanics of Erwin Schrödinger, we have to talk about the French physicist

Louis de Broglie (1892-1987). In his doctoral thesis, which he completed in 1924, he made a bold proposal. As explained in the penultimate section, wave-particle dualism was a characteristic exclusively of electromagnetic radiation. Why, according to de Broglie, this could not also be applied to matter? That is, why should matter not also have a wavelike character in addition to its real particle character? The examining board at the famous Sorbonne University in Paris was unsure whether they could approve it, so they asked Einstein. He was deeply impressed, so de Broglie got his doctorate. However, he was not able to present any elaborate description for the matter waves theory. Erwin Schrödinger then succeeded. In 1926, he introduced the equation that would be named after him. The circumstances surrounding its discovery were unusual. It is said that Schrödinger discovered it in late 1925 in Areas, where he was with his lover.

The Schrödinger equation is at the center of wave mechanics. As already stated, it is mathematically equivalent to Heisenberg's matrix mechanics, but it is preferred because it is much more user-friendly. There is a third version, more abstract, developed by the English physicist Paul Dirac. The three versions together form the non-relativistic quantum theory called quantum mechanics. As you may well have suspected, there is also a relativistic version.

The Schrödinger equation is not an ordinary wave equation, as it is used, for example, to describe water or sound waves. Mathematically, it is very similar to a "real" wave equation. Schrödinger could not explain why they were not identical. He had developed it by intuition. Based on the motto: What would a wave equation for electrons look like? This can also be called creativity. Very often, in the history of quantum theory, there was no rigorous derivation. It was trial and error until the equations that produced the desired result were found. Strangely, a theory of such precision could emerge from this. However, the theory as has a number of problems that have yet to be solved.

The solutions of the Schrödinger equation are the so-called wave functions. Only with them could the stability of atoms could be convincingly explained.

Quantum Origins of the Universe

Before trying to understand the dizzying quantum universe, it is necessary to familiarize oneself with the preceding scientific theories, with the classical physics. This body of knowledge is the culmination of centuries of research, begun even before Galileo's time and completed by geniuses such as Isaac Newton, Michael Faraday, James Clerk Maxwell, Heinrich Hertz and many others. Classical physics, which reigned unchallenged until the early 20th century, is based on the idea of a clockwork universe: ordered, predictable, governed by causal laws.

To have an example of a counterintuitive idea, let us take our Earth, which from our typical point of view appears solid, immutable, eternal. We can balance a tray full of cups of coffee without spilling a drop, yet our planet spins rapidly on itself. All the objects on its surface, far from being at rest, rotate with it like the passengers on a colossal carousel. At the Equator, the Earth speeds faster than a jet, at more than 1,600 kilometers per hour. Even more, it races wildly around the Sun at an incredible average speed of 108,000 kilometers per hour. On top of that, the entire solar system, including the Earth, travels the galaxy at even higher speeds. However, we do not realize it or feel like we are running. We see the Sun rising in the east and setting in the west, and nothing more. How is this possible? Writing a letter while riding a horse or driving a car at 100 km/h on the highway is a very difficult task, yet we have all seen footage of astronauts doing precision work inside an orbiting station launched around our planet at almost 30,000 kilometers per hour. If it were not for the blue globe in the background changing shape, it would seem that those men floating in space were standing still and not moving at all.

Intuition, in general, does not notice whether what surrounds us is moving at the same speed as us or whether this motion is uniform and not accelerated. We do not feel any sensation of displacement. The Greeks believed that there was a state of absolute rest, relative to the surface of the Earth. Galileo questioned this venerable Aristotelian idea and replaced it with a more scientific one: for physics there is no difference between standing still and moving with constant (even approximate) direction and speed. From their point of view, astronauts are standing still; seen from Earth, they are circling us at a crazy speed of 28, 800 kilometers per hour.

Galileo's sharpened ingenuity easily understood that two bodies of different weights fall at the same speed and reach the ground at the same time. For almost all his contemporaries, however, it was anything but obvious, because daily experience seemed to say otherwise. But the scientist did the right experiments to prove his thesis and also found a rational justification: it was the resistance of the air that shuffled the cards. For Galileo this was only a complicating factor, hiding the deep underlying simplicity of natural laws. With no air between the feet, all bodies fall with the same speed, from a feather to a colossal rock.

It was then discovered that the gravitational attraction of the Earth, which is a force, depends on the mass of the falling object, where mass is a measure of the amount of matter contained in the object itself.

The weight, instead, is the force exerted by gravity on bodies endowed with mass. You may remember what your physics teacher from high school always repeated to you: "If you transport an object to the Moon, its mass remains the same, while its weight is reduced." Today all this is clear to us thanks to the work of men like Galileo. The force of gravity is directly proportional to the mass: you double the mass, and the force is also doubled. At the same time, however, as the mass grows, so does the resistance to change the state of motion. These two equal and opposite effects cancel each other out and therefore all bodies fall to the ground with the same speed, not taking into account that complicating factor of friction.

To the philosophers of ancient Greece, the state of rest obviously seemed the most natural for bodies, to which they all tended. If we kick a ball, sooner or later it stops; if we run out of fuel in a car, it also stops; the same happens to a disc sliding on a table. All this is perfectly sensible and also perfectly Aristotelian. This Aristotelianism must be our innate instinct.

Galileo had deeper ideas. He realized, in fact, that by hinging the surface of the table and smoothing the puck, it would continue to run for a much longer time. We can verify this, for example, by sliding a field hockey puck over an icy lake. Let us remove all friction and other complicating factors and see that the puck continues to slide infinitely along a straight trajectory at uniform speed. This is what causes the end of the motion, Galileo said: the friction between puck and table (or between car and road) is the complicating factor.

Usually in the physics labs there is a long metal rail with numerous small holes through which air passes. This way a trolley placed on the rail, the equivalent of our disk, can move floating on an air cushion. At the ends of the track there are rubber buffers. A small initial push is enough, and the trolley starts bouncing non-stop between the two ends, back and forth, sometimes for a whole hour. It seems animated with a life of its own: how is it possible? The show is amusing because it goes against common sense, but in reality, it is a manifestation of a profound principle of physics, which is demonstrated when we remove the complication of friction. Thanks to less technological but equally enlightening experiments, Galileo discovered a new law of nature, which reads: "An isolated body in motion maintains its state of motion forever." By "isolated" we mean that no friction, other forces, or anything else is acting on it. Only the application of a force can change a state of motion.

It is counterintuitive, isn't it? Yes, because it is very difficult to imagine a truly isolated body, a mythological creature that you do not find at home, in the park or anywhere else on Earth. We can only encounter this ideal situation in the laboratory, with equipment designed as needed. But after witnessing some other version of the air track experiment, first year physics students usually end up taking the principle for granted.

The scientific method implies a careful observation of the world. One of the cornerstones of its success in the last four centuries has been its ability to create abstract models, to refer to an ideal universe in our minds, devoid of the complications of the real one, where we can hunt down the laws of nature. Having achieved a result in this world, we can go on the attack of the other, the more complicated one, after having quantified complication factors such as friction.

Let us move on to another important example. The solar system is really intricate. There is a big star in the center, the Sun, and there are nine (or rather eight, after the downgrading of Pluto) smaller bodies of various masses around it; the planets in turn may have satellites. All these bodies attract each other and move according to a complex choreography. To simplify the situation, Newton reduced everything to an ideal model: a single star and a single planet. How would these two bodies behave?

This research method is called "reductionism." Take a complex system with eight planets and the Sun, and consider a more tractable subset: one planet and the Sun. Now maybe the problem can be dealt with . Solve it and try to understand what characteristics of the solution are preserved when we go back to the starting complex system. In this case we see that each planet behaves practically as if it were alone, with minimal corrections due to the attraction between the planets themselves.

Reductionism is not always applicable and does not always work. That is why we still do not have a precise description of objects such as tornadoes or the turbulent flow of a fluid, or many complex phenomena at the level of molecules and living organisms. The method proves useful when the ideal model does not deviate too much from its messy and chaotic version, the one we live in. In the case of the solar system, the mass of our star is so much greater than that of the planets, that it is possible to overlook the attraction of Mars, Venus, Jupiter and company when we study the motions of the Earth: the star + planet system provides an acceptable description of the movements of the Earth. As we become familiar with this method, we can go back to the real world and make an extra effort to try to account for the next complication factors in order of importance.

The parabola and the pendulum
Classical physics, or pre-quantitative physics, is based on two cornerstones. The first is the Galilean-

Newtonian mechanics, invented in the seventeenth century. The second is given by the laws of electricity, magnetism and optics, discovered in the nineteenth century by a group of scientists whose names all were given to units of physical magnitude: Coulomb, Ørsted, Ohm, Ampère, Faraday and Maxwell. Let us start with Newton's masterpiece, the continuation of the work of our hero Galileo.

The bodies fall in free fall, with a speed that increases with the passage of time by a fixed value (the rate of change of the speed is called acceleration). A bullet, a tennis ball, a cannonball, all describe in their motion an arc of supreme mathematical elegance, tracing a curve called a parabola. A pendulum, that is, a body tied to a hanging wire (like a swing made by a tire tied to a branch, or an old clock) oscillates with a remarkable regularity, so that with it you can adjust your clock. The Sun and the Moon attract the waters of the terrestrial seas and create tides. These and other phenomena can be rationally explained by Newton's laws of motion.

Newton's creative explosion, which has few equals in the history of human thought, led him in a short time to two great discoveries. In order to describe them precisely and compare his predictions with the data, he used a particular mathematical language called infinitesimal calculus, which for the most part he had to invent from scratch. The first discovery, usually labelled "the three laws of motion," is used to calculate the motions of bodies once the forces acting on them are known. Newton could have boasted: "Give me the forces and a computer powerful enough and I'll tell you what is going to happen in the future."

The forces acting on a body can be exercised in a thousand ways: through ropes, sticks, human muscle, wind, water pressure, magnets and so on. A particular natural force, gravity, was at the center of Newton's second great discovery. By describing the phenomenon with an equation of astonishing simplicity, he established that all objects endowed with mass attract each other, and that the value of the force of attraction decreases as the distance between them increases. This way: if the distance doubles, the force is reduced by a quarter; if it triples, by a ninth; and so on. It is the famous "law of the inverse of the square," thanks to which we know that we can reduce the value of the force of gravity at will, simply by moving away far enough. For example, the attraction exerted on a human being by Alpha Centauri, one of the nearest stars (only four light years from us), is equal to one ten thousandth of a billionth, that is 10^{-13}, of that exerted by the Earth. If we approach an object of great mass, like a neutron star, the resulting gravity force would crush us to the size of an atomic nucleus. Newton's laws describe the action of gravity on everything: apples falling from trees, bullets, pendulums, and other objects located on the Earth's surface, where almost all of us spend our existence. But they also apply in the vastness of space, including between the Earth and the Sun, which are on average 150 million kilometers apart.

Are we sure, however, that these laws are still valid outside our planet? A theory is valid if it provides values according to the experimental data, taking into account the inevitable measurement errors. For example, the evidence shows that Newton's laws work well in the solar system. With a very good approximation, the individual planets can be studied in accordance with the simplification seen above, i.e. ignoring the effects of the others and considering the Sun alone. Newtonian theory predicts that the planets rotate around our star following perfectly elliptical orbits. But examining the data closely, we realize that there are small discrepancies in the case of Mars, whose orbit is not exactly the one provided by the "two bodies" approximation.

When studying the Sun-Mars system, we ignore the relatively small effects on the red planet from bodies like Earth, Venus, Jupiter and so on. The latter, in particular, is very large and gives some nice bangs to Mars every time their orbits come close. In the long run, these effects add up. It is not excluded that in a few billion years Mars will be kicked out of the solar system like a competitor out of a reality show. In other words, we realize that the problem of planetary motions becomes more complex if we consider long time intervals. Thanks to modern computers we can cope with these small (and not so small) perturbations, including those due to Einstein's theory of general relativity, which is the modern version of Newtonian gravitation. With the proper corrections, we see that the theory is always in perfect agreement with the experimental data. However, what can we say when even greater distances come into play, such as those between stars? The most modern astronomical measurements tell us that the force of gravity is present throughout the cosmos and, as far as we know, is valid everywhere.

Let us take a moment to look at a list of phenomena that take place according to Newtonian laws. Apples fall down from trees, heading towards the center of the Earth. Artillery bullets sow destruction following parabola arches. The Moon looms only 384,000 kilometers from us and causes both tides and romantic languor. The planets whizz around the Sun along orbits that are not very elliptical, almost circular. Comets, on the other hand, follow very elliptical trajectories and take hundreds or thousands of years to turn around and show themselves again. From the smallest to the largest, the ingredients of the universe behave in perfectly predictable ways, following the laws discovered by Sir Isaac.

Classical Physics vs. Quantum Physics

Physicists are trying to understand the world, but to date no proven and palpable theory has emerged to bring the worlds of classical physics and quantum physics together, and help us ultimately to understand where we come from, where we are, and where we are going. At the end of the day, these are the fundamental questions posed by both classical and quantum physics.

In classical physics, as drawn out by Einstein's principle of general relativity, reality is made out of 4 dimensions, also called the space-time continuum. In this paradigm, gravitational fields are continuous entities. In quantum mechanics, however, fields are not continuous, but discontinuous. They are not defined by the 4 dimensions but by "quanta." As such, concepts like the "gravitational field" are missing from the world of quantum physics, which is also the biggest bridge that classical physicists and quantum researchers have to build between their points of view.

This is not just a matter of fancy definitions. The world of quantum mechanics and the world of classical physics are incompatible because they describe reality in completely different ways, in different terms, and from different perspectives that never meet at any point. In classical physics, things happen for a reason. They take place according to the old cause-and-effect dictum.

Nothing happens randomly, it only happens because there is something earlier that has caused it. In quantum physics, scientists do not see reality in terms of cause and effect, but in terms of particles jumping from one state to another based on probability, rather than on definite outcomes.

Why is reconciliation important, especially given that these two disciplines seem so different and at such a deep level?

It is important because reconciliation would create a whole new theory that would explain the universe on both a small and a large scale. Where classical physics fails to give explanations of the microcosm, quantum physics would succeed. Where quantum physics fails to make sense when it is blown up to macro-objects, such as the infamous cat that was both dead and alive, classical physics would be able to breathe in some logic.

Fundamentals of Classical Physics
Classical physics is invariant for time-reversal and we have seen that this gives us serious problems when we try to find an explanation for the thermodynamic arrow of time. It is therefore important to investigate the dynamic reversibility in quantum mechanics.

Before we take a look at particle physics, quantum theory, and the cosmos, we need a brief introduction to the concepts of classical physics:

- Energy
- Weight and mass
- Matter - solids, and liquids
- Measurements and units

Energy

Energy must be transferred to an object to do work on or to heat it. Newton's law of the conservation of energy states that energy may be transformed from one form to another. It cannot be created nor destroyed. The SI unit for energy is the joule.

Types of energy include:

- Kinetic energy (movement)
- Chemical energy (e.g. coal, natural gas, etc.)
- Thermal energy
- Magnetic energy
- Light energy
- Electric energy
- Gravitational potential energy
- Nuclear energy

Weight and Mass

Weight is therefore dependent on the gravitational force where the object is situated. According to the International System of Units (SI Units), weight is measured in Newtons with the symbol N.

Mass can be described simply as the amount of matter in an object. This measurement is given in kilograms or grams and is calculated by multiplying the object's volume by its density. The mass of an object will be the same no matter where it is measured, because the object will always contain the same number of protons, or amount of matter.

Matter

Matter. There are several forms of matter we know of, but only three are relevant to this book: solids, liquids, and gases.

- Solids. Solid objects have a defined shape because their atoms are packed tightly together (i.e. they have a high density). Their atoms cannot move around and cannot be compressed into a smaller volume.
- Liquids. In liquids, the atoms are not so tightly packed so they can flow around each other. Most liquids can be compressed into a smaller volume in a container, i.e. their atoms can be forced closer together (they can become denser).
- Gases. The atoms in gases are in constant movement and have a relatively large space between them. Gases can be compressed into a smaller volume when confined in a

container and they expand when released.

Measurements and Units

- Density. The density of matter is a measure of how closely its atoms are packed together. Density is measured in kilograms per cubic meter and can be calculated by dividing the mass of an object by its volume.
- Volume. It is the amount of space that matter occupies. There are many ways of measuring volume depending on whether you are measuring solids, liquids, or gases. The formulae for measuring different shapes of solids or containers of liquids or gases are different.

The units in which volume is measured can be confusing. The usual system in the US differs from the imperial system in the UK, and both of those differ from the metric system. In 1960, an international system of units was introduced. Le Système International d'Unités, now known simply as SI Units, is used by the scientific community to avoid confusions.

Fundamentals of Quantum Physics

The Quantization of Light

This was a forward step taken by Albert Einstein in 1905. With it, he suggested that quantization did not just involve mathematical tricks. He added that it also involved the beam of light energy that is in the individual packets, currently referred to as photons. The energy of a single photon will be given by the product of the frequency of the energy and Planck's constant.

In the 19th century, light was considered to be flowing in waves and this explained the behaviors of light such as polarization, diffraction and refraction. Actually, according to James Clerk Maxwell, magnetism, light and electricity were all manifested by the same phenomenon, which was the electromagnetic field. He explained light as waves that were a combination of magnetic fields as well as oscillating electricity. Einstein's "photon model" came into place when it was able to successfully explain the photoelectric effect. This effect is explained in the next section.

The Photoelectric Effect
According to Einstein's explanation, he argued that a beam of light has photons, which are particle streams, and also has a frequency "f". The energy present in that photon will be equal to "hf". This means that there is no effect on the energy that relates to the beam's intensity. He further explained that to remove an electron from a given metal, a "work function" is required, which is a certain energy amount denoted by "φ". With his further explanation, if the work function is higher than the photon's energy there will be no sufficient energy to remove the electron from the given metal.

His description also argued that light is composed of particles that gave an extension to Planck's notion. This is the notion that energy is quantized, whereby more or less amount of energy can be delivered by a given photon depending on its frequency. This was a compromise on the particle state of light since it explained that light also had waves. As a consequence, this resulted in the theory of the quantization of light.

Matter Quantization

According to Bohr, electrons jumped from one orbit to another. In that case, they gave off the light emitted in the form of photons. The energies that were emitted by the photons were highly dependent on the differences in energy between the orbits. First, there were very many criticisms of Bohr's model. Many argued that this model was wrong; however, in the end, it was evident that the model was good to suit quantum physics. With his explanation, Bohr argued that matter also had some wave-like properties. According to him, an electron beam can also exhibit "diffraction." This is a case similar to a beam of light or a wave of water. Thus, small molecules and atoms show the same phenomenon. To prove the above, a double-slit experiment was undertaken. This experiment is explained in the next section.

Time in Quantum Physics

Time is used in quantum theory as an external concept. We can say that, like in classical physics, its role is that of controlling motion, either as absolute time in ordinary quantum mechanics or as proper time in a classical spacetime metric, as in the case of quantum field theory. As a consequence, the parameter t, which appears in Schrödinger's equation, can be identified with Newtonian absolute time. Unlike space variables, it is not an operator. When we turn to quantum field theory, which includes special relativity, the wave function becomes a function of the quantum fields and of time, i.e.:

$$\Psi = \Psi(\Phi_1(x, y, z), \Phi_n(x, y, z), t)$$

In this case, the operators are the fields, while the space coordinates are only indices. As we can see, the dependence on time is separated from the dependence on the fields. The theory is now based on Minkowski spacetime.

Although time appears in quantum theory as Newtonian time or as Einsteinian proper time, it nonetheless shows some very peculiar features. To start with, the notion of the past changes radically when we go from classical to quantum mechanics. In quantum physics, as we have seen on the double-slit experiment, the past before a measurement seems to be embedded in a fog. Even if we "now" measure an electron at each position in the past, i.e. before this measurement, we cannot say anything about the position of the electron. We cannot even imagine the electron to be in a given position, as we know that the wave function, representing its state, embraces all possible positions.

The Delayed-Choice Experiment

The peculiar features of time in quantum physics are clearly seen in a thought experiment proposed by John Wheeler (1983), which is known as the delayed-choice experiment. The experiment wants to show if it is possible for one to choose if a photon should behave as a particle or as a wave, and possibly change its "choice."

The experiment is based on an interferometer, shown in Figure 4.1. S is a source of photons. The paths of the photons in the beam are separated by an optical device called a beam splitter (BS1, see Figure 4.1a). It is just a half reflecting mirror capable of separating a beam into a reflected (R) and a transmitted part (T). For an ideal beam splitter, each of the beams R and T will have an intensity which is 50% of that of the initial beam. Now the two separated beams are reflected by two identical mirrors M, and then collected in two detectors (D1 and D2). Each detector will measure a beam's intensity, which is half that of the initial beam.

Suppose now that we insert a second beam splitter (BS2) in the position indicated in Figure 4.1b, such that both rays deflected from the mirrors M are incident on this beam splitter. With this addition, the two detectors D1 and D2 will see something different from what was obtained in the configuration of Figure 4.1a. Each of the rays coming from the M deflections will be partly reflected and partly transmitted by BS2.

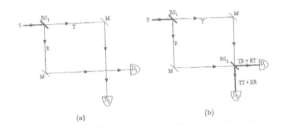

Figure 4.1: Scheme of the delayed-choice experiment

Then, in one of the detectors (for example, D1) the two rays will superimpose with a destructive interference so that this detector will give no signal. In the same case, however, the rays will interfere constructively before reaching the second detector, so that it will register a signal and the intensity of this signal will be equal to that of the original beam. Thus, by inserting or not inserting the second beam splitter, we alter the results of the experiment.

To emphasize the weirdness of what we have seen, let us now reason in terms of single photons and suppose that the intensity of the source S can be reduced to the point that only single photons interact with the first beam splitter. We should now say that each photon has a 50% probability of being reflected or transmitted and, therefore, of following either path R or path T. In the end, and for the configuration of Figure 4.1a, we expect that each photon will be detected by one or the other detector and with equal probability.

When we insert the second beam splitter (see Figure 4.1b), the photons behave as waves and have the possibility of interfering, like the electrons of the double-slit experiment. We then expect a signal from the detector, which is aligned with a constructive interference, and no signal for the other detector that will not be reached by any photons.

In the first case of only one beam splitter, each photon behaves as a particle, while in the second case, with two beam splitters, each photon behaves as a wave. Obviously, we could also choose, during the experiment and before the measurement, to remove the second beam splitter, thus constraining the photons to behave as particles. The weird thing is there is no way in which the photons would know our choice.

We therefore arrive at an apparent paradox. After their arrival to the first beam splitter, the photons do not know if there is a second beam splitter. And yet, at the end, they behave as if they knew of its presence or its absence.

The weirdness increases if we delay our choice of using BS2 or not, and we make that choice after the interaction with the first beam splitter when the photon is in the middle of the interferometer's arms.

Wheeler has emphasized the paradox of the delayed choice experiment by imagining the same experiment on a cosmic scale (see Figure 4.2).

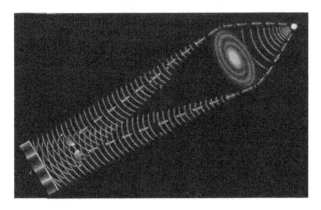

Figure 4.2: A galaxy acts like a gravitational lens for the light emitted by a quasar.

In this case, the source S is not in the laboratory but rather a quasar billions of light years away from Earth. A galaxy on the way from the quasar source to Earth works as a gravitational lens (an effect predicted by Einstein) and forces the light rays to follow two different routes on each side of the galaxy to reach Earth. In other words, the galaxy has the same effect as the BS1 beam splitter from the laboratory experiment.

An observer on Earth would then collect the two beams and orient them towards a real beam splitter, followed by two detectors. Also, in this hypothetical experiment, we should expect a figure of interference from one detector and no signal from the other. In case the beam splitter on Earth were removed, we would expect the photons to be registered by one detector and the other without any sign of interference.

It is clear that, in this cosmic experiment, the delayed choice could be something drastic. The photons would have traveled for billions of years along only one or both of the possible paths around the galaxy, without knowing whether a beam splitter was present on Earth or not. The beam splitter could be inserted or removed by the experimenter to obtain results according to a given experimental choice.

It would therefore seem that our current actions, actions that we do now, determine the behavior of photons emitted billions of light years ago from the quasar. This would imply a violation of the principle of causality.

However, this conclusion can be avoided. The paradox that our present actions determine something from the past is due to our classical (i.e., non-quantum) perception of the world. It is on this basis that we believe that the photons either follow a unique path or follow both. This view, in quantum terms, is false. As we have said, the quantum past is embedded in a fog.

Photons do not do one thing or another. They are instead in the constant superposition of different alternatives and it is by the act of measurement that only one of them materializes. We do not have to think that some billions of light years ago the photons decided what to do—they have always been in a quantum reality that is a sum of all possibilities.

In fact, similar experiments to Wheeler's thought have been done. Although their weirdness is explained by quantum theory, they nonetheless seem to mix past and present actions, in apparent contrast with the principle of relativity. In any case, they indicate the very peculiar role of time in quantum mechanics.

Dynamic Reversibility in Quantum Mechanics

Classical physics is invariant with respect to time reversal, and this gives us serious problems when we try to find an explanation for the thermodynamic arrow of time. It is therefore important to discuss dynamic reversibility in quantum mechanics.

We have to recall once again that there are two different aspects of quantum mechanics. The first, which we will call the dynamical aspect, is represented by Schrödinger's equation, which gives the deterministic evolution of the wave function for an undisturbed system.

The second aspect pertains to the process of measurement or, more generally, to any interaction process. What happens there, according to the given rules of quantum mechanics, is not described by Schrödinger's equation. If we stick to these rules, what happens is completely casual, i.e., not deterministic, and irreversible in time.

When talking about the reversibility of time in quantum mechanics, we now want to refer to the first of the two previous aspects, i.e. to the evolution described by Schrödinger's equation.

Let us consider an operation of time reversal characterized by the transformation:

t→-t   p→-p   x→x

Under this, a state vector $|\Psi\rangle$ will transform into some other state vector $|\Phi\rangle$. Now, to test the invariance of quantum mechanics with respect to time reversal, we will have to ask about the invariance of the scalar products (which represent probabilities), i.e.

$|\langle\Psi'|\Phi'\rangle| = |\langle\Psi|\Phi\rangle|$

and also, the invariance of the expected value of any variable, indicating with M the operator corresponding to that variable,

$|\langle\Psi'|\Phi'\rangle| = |\langle\Psi|M|\Phi\rangle|$

As long as the evolution of time of the state vectors or wave functions is governed by Schrödinger's equation, we must turn to that equation to verify the above equalities. Now, by changing t into −t, and Ψ into Ψ', Schrödinger's equation changes to

$i\hbar\, \partial |\Psi'\rangle/\partial t = - H\, |\Psi'\rangle$

which is a different equation from that for Ψ. However, if we choose

$\Psi'(t) = \Psi^*(-t)$

where the * denotes the complex conjugate, we see that we recover for Ψ' the same equation we had for Ψ. The invariance with respect to time reversal of Schrödinger's equation implies that, in addition to inverting time, we also have to change the wave function into its complex conjugate.

It can be seen, though we are not giving the details of the proof, that the same choice of Ψ' also ensures the invariance of the scalar products and of the expected value of any variable.

Should we conclude from this that quantum mechanics is invariant with respect to time reversal? We believe that the answer is negative, since this conclusion is based only on Schrödinger's equation, and we know that this is not the whole story. Schrödinger's equation only holds when a quantum system is unperturbed, between measurements, or between interactions with the external world.

Rutherford's Experiment

In 1911, Rutherford and his partners Hans Geiger and Ernest Marsden started a progression of tests that would totally change the celebrated particle model. They bombarded little sheets of gold with fast moving alpha particles.

Since according to the affirmed atomic model the size and charge of atoms were consistently disseminated all through their space, the scientists expected that alpha particles would go through the gold sheets with almost no deflection. However, some were even redirected back to the source. No prior data had prepared them for this result. In his aphorism, Rutherford proclaimed, "it resembled shooting a 15-inch shell at a piece of tissue paper, and it came back and hit you."

Rutherford needed to come with an entirely new atomic model to explain his results. Since, by far, the majority of the alpha particles had passed through the gold, he reasoned that most of the atom was empty space. He imagined that all the positive charge and size of the particle mass ought to be situated in the littlest space inside, which he called the nucleus. The nucleus was thus a little thick molecule comprised of protons and neutrons.

Rutherford's atomic model is commonly known as the nuclear model. In the nuclear atomic model, protons, and neutrons, which comprise nearly all the mass of the atom, are located in the nucleus at the center of the atom. Electrons are distributed around the center and occupy most of the volume of the atom. It merits stressing how small the nucleus is compared to the rest of the atom. If we could blow up an atom to the size of a large football field, the nucleus would be about the size of a marble. Rutherford's model, in the end, turned out to be a critical improvement in the full comprehension of the atom. However, it did not completely address the nature of the electrons and the way they occupied the immense space around the nucleus. Rutherford was granted the Nobel Prize in Chemistry in 1908.

Wave-Particle Double-Behavior
The study of realism tries to explain rules, as it focuses on improvement and matter. Regardless, quantum material science is trying to understand the structure of minuscule particles and how they move. Such particles contain components, including electrons, protons, and neutrons.

In its emphasis on non-perpetual particles, the study of materials determines the particles that make up little particles. Rules managing outwardly debilitated structures have plainly been mistaken in deciding the area of low-lying zones since the start of the 20th century. "Quantum" starts with the Latin word for "esteem." Material science is utilized to refer to small units of results, and the life of their work is normal and found in quantum physical science.

Altogether, even the most insistent and reformist circumstances, for instance, are practical, despite the fact that they appear to be small.

The quantum model of particles is significantly more befuddling than what we have seen previously, instead of spinning around a star-like center, electrons, and hovering in an undetectable, lesser-known, or cloud-like turn of events. The last arrangement we have in the electron assortment, referring to the number of external shell electrons, is frequently linked to a fixed request.

This brings us to understand the quantum idea of light to mirror the logical name of the material, to understand that its thought process should show the capability of the electron space at some irregular time. Along these lines, when the word is related to "light of light," you should have a solid comprehension of the overall guideline of the law.

The likelihood that contemplating anything can impact the physical cycles that happen is not the same as the study of realism. For instance, in what is known as wave-atom duality, light waves move like particles, and these particles also move as waves. Put another way, light has the qualities of particles and waves.

In the quantum upgrade, the issue can move between various places without moving from space to both. This gives away the current application where the information can be isolated by a different split. As per quantum science, we find that the tone of the universe can be referred to as a continuation of possibility.

There are numerous types of materials science. The one that centers for the most part on the conduct of light (photons) is known as Quantum Optics.

When researching Quantum Optics, you will find that instances of the improvement of individual photons (light bars) straightforwardly influence light. An essential and adaptable instrument known as a laser is one of the numerous critical tools in Quantum Optics.

This is a distinct difference from Sir Isaac Newton's larger light examination. Classical Optics, where light seemed to have only sub-atomic structures, implying that it moved in a methodical manner, returning to objects, and passing through items with immaterial obstructions.

Photons

To make it clearer what is meant when the word photon is used, let us guide our focus toward the Photon Theory of Light. Photon is a canny (or quantum) parcel of electrical vitality in this specific sense.

Placed on a clear and stable machine, photons have a consistent light speed for all watchers. It happens at the speed of light, C, where:

$C = 2,998 \times 10^8$ m/s height

Altered in the Photon Theory of Light, the main attributes of photons are as follows:

- They move at a consistent motion, $c = 2.9979 \times 10^8$ m/s (low speed) in free space.

- They are known to have zero trouble and eagerness.

- They communicate vitality, contrasted with the recurrence of nu and recurrence of lambda, (and p, vitality) of an electromagnetic wave, by

$E = h\nu$ ye no $p = h/lambda$

- They can have particle-like interactions, for instance electron impacts and other transitory sections.

The more we understand the quantum properties of light, the more we can incorporate part of the relating cycles (maintenance, yield, and vivified yield) into the laser, as this is one of the most surprising uses of quantum optics. For the most part, these three can be summarized in various light sources at expanding levels.

Electronic advances are normally a sort of progress that communicates or consolidates light. Imagine an electron moving between atomic size levels to perceive how this works.

For a laser to work properly, vivified light results are fundamental. Sustainable light yield is utilized to give the advancement expected to play out a basic reasoning capacity.

The tale property known as level headedness is the aftereffect of an amended leave rate. Ordinary advancement makes discharge times that are needed to give improved light. This fixes the delivered photons in a decent pre-request plan where all the photons have a full stage relationship to one another.

This sort of insight (related arranging) is described by two unmistakable terms: impermanent affectability and area mindfulness. Both wind up being critical in the advancement of the blockchain used to produce perceivability.

De Broglie Hypothesis
This was one of the most acclaimed logical meetings ever. Of the 29 up-and-comers, 17 received Nobel prizes. The gathering is significant for two titans of material science: Niels Bohr and Albert Einstein.

1927 was the year, and researchers were stunned. Are electrons, lights, and comparable items waves or particles? In certain tests, the little bodies acted like waves, and in others they acted like particles. This is not going on in our large world. Sound waves do not act like rocks, which is fortunate, since otherwise your ears would suffer.

The 1927 Quantum Mechanics meeting discussed a combination of terms that appeared to be in conflict. Schrödinger and de Broglie introduced their perspectives. Bohr had his own opinion. It was later called the Copenhagen Interpretation. The agreement, following Bohr's recommendation, was that wave estimations were characterized as materials, for example, electrons. However, as particles, they did not exist until somebody needed them. Utilizing Bohr's own words, these elements had no "obvious life in the typical setting." None of that would have been Einstein.

In fact, Einstein would not fully agree. An electron was an electron, and even if nobody was looking at it, it was still there. Towards the end of the meeting, Einstein tested Bohr's view. That was just the start. They were close, and communicated often until Einstein's death, both in person and in writing.

Their discussions were those of courteous partners. Bohr and Einstein were old buddies and regarded each other fondly. Notwithstanding, they persevered each one in their opinions. Bohr said, "It's not reasonable to believe that material science needs to discover what nature resembles." Einstein did not agree with that idea. "The main reason why we are scientists is to discover what it is," he would say.

For all its unpredictability, Bohr's Copenhagen Interpretation continues to be one of the world's most broadly acknowledged quantum material science ideas.

Numerous definitions of it seem mostly strange, but they all highlight one straightforward truth. Our universe is a secret, as all researchers will let you know. It befuddles us with unfathomable realities. Possibly one day we will understand them in more detail. However, before that we should confront the great secrets around us.

On the other hand, Planck's time is the fundamental unit of time in the arrangement of Planck Units, meaning:

$t_p = 5.39 \times 10^{-44}$ s

In SI units, time estimations are made quickly, aside from the fact that the utilization of seconds has the benefit of an everyday presence. For instance, estimating the time it takes for a contender to run 100 meters is very little on the planet. However, it is different when we talk about other occasions in the early universe. We have to look, for instance, to what happened in the 10^{-35} seconds after the Big Bang.

The consequence of using seconds to quantify time is that big changes are expressed in figures that are not easily recalled:

Light speed $c = 299,792,458$ m/s (s)

Gravity $G = 6.673 (10) \times 10^{-11}$ m3 kg (- 1) s (- 2)

Board Strength (diminished) $\hbar = h/2\pi = 1.054571596$ (82) $\times 10^{-34}$ kg m2 s-1

Boltzmann strong k = 1.3806502 (24) x 10-23kg m2 s-2K-1

Planck's time uses a mix of these key components:

By revising the base units of length, size, and time with Planck units, the main points of interest are:

$c = G = \hbar = k = 1$

Actually, Planck's time is the time it takes for a photon to travel Planck's length:

$= 1.62 \times 10^{-35}$ m

This is the briefest time limit conceivable. With its overall length of Planck, Planck's time characterizes the scale at which the current assemblage of thought is presented.

Accordingly, our present introduction of the main universe improvement starts at $t_p = 5.39 \times 10^{-44}$ seconds after the Big Bang.

Quantum Tunnelling

Tunneling is a function of mechanical quantity. A tunnel occurs when the electrons in it move across a barrier that they would not move in the classical way. In classical words, if you have no energy to "overcome" an obstacle, you will not. In the quantum mechanical world, electrons have wave-like properties. Such waves do not stop suddenly, but gradually taper off at a wall or barrier. If the barrier is relatively thin, the probability function will enter the next area via the barrier. Because an electron is low on the other side of the barrier, other electrons are traveling and emerging on the other side. It is called tunneling when an electron passes this way through the barrier.

Quantum physics tells us that electrons have wave-like properties and particle-like characteristics. Tunneling is a wave-like result. When an electron wave hits a boundary, it does not stop suddenly, but it tapers rapidly and exponentially. If the barrier is thick, the wave will not pass. Part of the wave passes, and some electrons may pass to the other side of the barrier.

The number of electrons that pass the tunnel depends on the width of the tube in which the body is loaded. The sum of electrons pumped into the tunnel is highly dependent on the size of the tunnel in the tunnel pipe.

To extend this full description to the STM: the main starting point of the electron is either the sample or the tip, which depends on the instrument's setup. The boundary is the distance (vacuum, air, liquid), depending on the experimental setup, the second area is the reverse, i.e., sample or edge. The size of the current is measured by measuring the current through the distance.

Piezo-Electric Effect

In 1880, Pierre Curie discovered this effect. It is the resulting electric charge that accumulates in crystals such as quartz or titanium barium and other materials, in response to crushing or mechanical stress. The effect is that opposite charges are generated on the sides. These components are used to scan the tips in microscopic scanning tunneling (STM) and other scanning research techniques. Lead zirconium titanate (PZT) is a typical piezoelectric substance used in the scanning method microscopy.

The Black Body Radiation

For the time being, we will go to another riddle that disdained researchers as the new century turned (1900): how do warm bodies start? There was a complete understanding of the involved framework. Heat was known to make particles vibrate with enthusiasm, and particles and atoms then showed instances of electrical charges. Clearly, Newton was fit as a fiddle. However, from examination by Hertz and others, Maxwell's possibilities for light-emanating redirection cases were confirmed. From Maxwell's conditions it was known that the radiation went at the speed of light, and for this situation, it was perceived that the light itself, alongside the warm beams related to the field, were actually electric waves. At that point, the picture was that when the body was heated, the resulting vibrations on the sub-nuclear and atomic-scale were inevitably eliminated—recognizing at the time that Maxwell's concept of electromagnetism, which is the most effective in the physical world, was genuine. At the sub-nuclear level, these attractive conditions would have passed, maybe radiating warmth and perceptible light.

How is Radiation Absorbed?
What is implied by the term "dull body"? The truth is that the radiation of a hot body assumes that the body is being warmed. We ought to consider how various materials store radiation to see this incredible

achievement. For example, an item appears to get light in any capacity, and light passes straightforwardly. With a sparkling metal surface, light is excluded.

How might we comprehend these different cycles, such as light waves that adjust to changes in applications, making these charges influence and store heat from radiation? On account of the glass, unmistakably, this is not going on.

How might we understand light on metal? A small piece of metal has electrons permitted to go. This is the thing that makes iron be a metal: it conducts vitality and warmth viably; the progression of these straightforward electrons really sends both.

Right now, what might be said about taking a look at something that focuses light: there is no transmission and no showcase. We are moving toward the best end with cinders.

Like steel, it will lead to the progression of power, be that as it may. There are detached electrons, which can go at all energies, yet keep on holding objects. They have a momentary significance. At the point when they thumped, they caused an upheaval, similar to the balls hitting the watchmen on a pinball machine, so they emitted a solid power in the warmth. Aside from the way that electrons in debris have a shorter length contrasted with that of metal, they are plainly contrasted with the electrons limited by particles (as in glass), so they can quicken and pick up vitality in the electric field. Along these lines, they are groundbreaking go-betweens in moving vitality from light waves to warm.

Absorption and Emission
For what reason do bodies move when they are warmed? The pinball machine's similitude is as yet worthy: think now about the pinball machine where the limits are, etc.

We can be more exact: the body emanates heat at a given temperature and returns similarly as it produces heat.

Kirchhoff has exhibited this. The essential point is that on the off chance that we believe that a specific body is heated, at that point in a room loaded with things of a similar temperature, it will include radiation from various bodies. This implies it will improve, and the remainder of the room will be cold, dismissing the second law of thermodynamics. We can utilize such a body to construct a warm vehicle that isolates the occupying as the room gets colder and colder.

Nonetheless, the metal sparkles when it is sufficiently warm: for what reason would it be so? As the temperature rises, the alternate way bit of the particles vibrates at a consistent level; this development scatters and accelerates electrons. Undoubtedly, even glass is lit up at temperatures sufficiently high, as electrons radiate and move.

The Absorption of the Radiation

Dark Body Spectrum

A body at any temperature over zero will emit radiation. The recurrence of radiation relies upon the specific body structure.

We ought to inspect how this works. In 1859, Kirchhoff had a shrewd thought: a little hole on a huge box is a great assurance, on the grounds that any beams that discover a hole hop around within, are held near each weave, and have minimal possibility of getting out.

Along these lines, we can do this for a change: we have a grill with a little hole as an afterthought, and possibly the radiation from the space is adequate to show the correct producer as we will discover. Kirchhoff has aggravated researchers and experimentalists to bode well and measure separately the bowing/horrible intensity of this "radiation," as he calls it. It was Kirchhoff's examination in 1859 that formally eliminated quantum theory forty years afterward.

Observations

By the 1890s, arraignment techniques had progressed to the point that they could not be viewed as an exact estimation of the radiation's greatness in an opening, what we would call dark radiation. In the last years of the 1800s, at the University of Berlin, Wien and Lummer twisted around the side of a totally shut oven and started to isolate the coming beams.

The locater was off the whole screen to discover how much style was communicated in each reiteration. This is a specialist case model; genuine test blueprints are exceptionally refined. For instance, to raise hell free infrared estimations, repeating waves are executed by various quartz signals and various qualities. They discovered rehashed radiation/twisting bends close to this (right):

The obvious range begins at about 4.3×1014 Hz, so this oven sparks a dull red.

One little point: this structure is the force of the power inside the oven, which demonstrates (f, T), which implies that at a temperature of T, the power of Joules/m3 in the repeat of straightforward f, f + δf is ρ (f, T) if.

To get the vitality out of the opening, recollect that the radiation inside the oven has similar waves that move in two unique ways - so half of them will come out through a hole. Also, if the hole has an A position, the waves come in during a period that will see the objective zone less. The aftereffect of these two impacts is as per the following.

Radiation Energy from the Gap Region
A = 14 A cρ (f, T)

They were set up to implement Stefan $P = \sigma T4$ and Wien's Transfer Law by restricting the way that the dark body twists at various temperatures.

Imagine a scenario where we look again, and this twists in detail: discovering low waves, f, (f, T) found to compare to f2, the spellbinding state, yet by expanding f, it falls underneath the parabola, fmax, at that point diminishes quickly towards zero as though the past fmax increment.

The curve ρ (f, 2T), then, is regularly the length of ρ (f, T). (See outline above.) It is also twice as wide as the even level, so the field beneath the shell, in correlation with the energetic surface, increments the temperature by multiple times: Stefan's law, $P = \sigma T4$.

Basic Laws

The essential supposition that depends on the radiation test perspective on the hole.

Stefan's Law (1879)

Complete P power from one square meter of the dark region in temperature T goes as one fourth of the absolute temperature:

$$P = \sigma T4, \sigma = 5.67 \times 10-8 \text{ watts/sq.m.} /K4$$

After five years, in 1884, Boltzmann found this T4 conduct in principle: he utilized conventional thermodynamic speculation for a situation stacked with electromagnetic radiation, using Maxwell's wonders to relate the force and power of vitality.

Wien Relocation Act (1893)

As the grill shifts temperature, so does the reiteration where the radiation is sent as often as possible. That is additionally legitimately identified with the general temperature:

$f_{max} \propto T$

Wien himself found this law by theory in 1893, after Boltzmann's hypothesis about thermodynamics. It had been watched by the American space expert Langley.

This skyscraper in fmax and T is normal for everybody. When the metal is warmed in a fire, the principle obvious beams (about 900K) are ruddy, almost no noticeable re-noticeable light. Further increment in T makes a hazier shade from orange yellow, in the long run, to blue and to higher temperatures (10,000K or higher) when high radiation presentation is plainly noticeable.

This is a dreary advance where the best power is significant in keeping up sun-related vitality. The glass should give the sun's beams access. This is justifiable because the two beams are at a totally extraordinary recurrence, 5700K and, state, 300K, and there are immediate to-light items that are wrong in infrared radiation. Kindergartens work in light of the fact that fmax changes with temperature.

Black Body Curve

These planned test outcomes are an approach to change. The essential trial of information was reviewed by Max Planck in 1900. He zeroed in on featuring the troublesome cases that should be available in oven partitions, which emerge from inner warmth and - in thermodynamic settings are themselves driven by the radiation field.

Essentially, he found out that he could speak to the watched bend on the likelihood that he needed these oscillators to show up as dynamic, as the good old view would ask. However, they could basically lose or take power with sections, called quanta, size hf, for augmentation oscillator f. The fixed h is at present called Planck's agreement, h = 6.626 × 10−34 joule/sec.

At that demand, Planck decided the extent compared to the greatness of the radiation inside the oven:

(f, T) df = 8πVf2dfc3hfehf/kT - 1

A superior comprehension of this formula with explicit tests and the ensuing requirements for imperativeness quantization turned into the most significant material science progression for a century.

Nobody saw it for long. His dark body twist was totally acknowledged as a right: a developing number of direct tests demonstrated it commonly; however, the outrageous idea of quantum thinking did not enter. Planck didn't stress excessively; he did not think it was conceivable. He accepted it as a unique amendment after some time, which would have appeared to be ridiculous.

A contributor to the issue was that Planck's excursion to the condition was long, strenuous, and unthinkable, even to the point of making opposing suppositions of different classes, as Einstein later called attention to it. In any case, the outcome was positive regardless, and to comprehend why we would follow another, easier, the course that was begun, yet not effectively finished, by King Rayleigh in England.

Light

Light is a form of energy. It can be produced in various ways, transforming electrical energy, as seen for example in a light bulb, or in the redness of toaster resistances, or chemical energy, as in candles and combustion processes. Sunlight, a consequence of the high temperatures present on the surface of our star, comes from the nuclear fusion processes that take place inside. The radioactive particles produced by a nuclear reactor here on Earth emit a blue light when they enter the water (which ionize, i.e., that is, tear electrons from atoms).

It only takes a small amount of energy to heat any substance. At small scales this can be felt as a moderate rise in temperature. As those dabbling with DIY on weekends know, nails get warm after a series of hammerings, or if they are torn from the wood with pliers. If we supply enough energy to a piece of iron, it begins to emit light radiation. It is initially reddish in color, then as the temperature increases, we see orange, yellow, green and blue tones appear in that order. In the end, if the heat is high enough, the emitted light turns white, the result of the sum of all colors.

Most of the bodies around us, however, are visible not because they emit light, but because they reflect it. Excluding the case of mirrors, the reflection is always imperfect, not total: a red object appears to us as such because it reflects only this component of the light and absorbs orange, green, violet, and so on. The pigments of paints are chemical substances that have the property of accurately reflecting certain colors, with a selective mechanism. White objects, on the other hand, reflect all the components of light, while black ones absorb them all: this is why the dark asphalt of a parking lot becomes hot on summer days, and this is the reason why in the tropics it is better to dress in light colored clothing. These phenomena of absorption, reflection, and heating, in relation to the different colors, have properties that can be measured and quantified by various scientific instruments.

Light is full of oddities. We see you because the light rays reflected from your body affect our eyes. How interesting! Instead, our mutual friend Edward is observing the piano: the rays of the you-us interaction, which are normally invisible, except when we are in a dusty or smoky room, intersect with those of the

Edward-piano interaction without any apparent interference. If we concentrate the beams produced by two flashlights on an object, we realize that the intensity of the light doubles, so there is interaction between the light rays.

Let us now examine the goldfish tank. We turn off the light in the room and turn on a flashlight. Helping ourselves with some dust suspended in the air, maybe produced by banging two blackboard erasers or a dust rag, we see that the light rays bend when they hit the water (and also that the poor little fish is looking at us perplexed, hopefully waiting for food). This phenomenon by which transparent substances such as glass deflect light is called refraction. When Boy Scouts light a fire by concentrating the sun's rays through a lens on a bit of dry wood, they are taking advantage of this property: the lens bends all the light rays, making them concentrate on a point called "fire," and this increases the amount of energy until it is so high that it triggers the combustion.

A glass prism is able to decompose light into its components, the so-called "spectrum." These correspond to the colors of the rainbow: red, orange, yellow, green, blue, indigo, and violet. To memorize the order, remember the initials ROYGBIV. Our eyes react to this type of light, called "visible," but we know that there are also invisible types. On one side of the spectrum, there is the so-called "infrared" long wave range. Of this type, for example, is the radiation produced by certain heaters, the toaster resistances or the embers of a dying fire. On the other side there are short wave "ultraviolet" rays. An example is the radiation emitted by an arc welding machine, and that is why those who use it must wear protective glasses. White light, therefore, is a mixture of various colors in equal parts. With special instruments, we could quantify the characteristics of each color band, more adequately their wavelength, and report the results on a graph. By subjecting any light source to this measurement, we find that the graph assumes a bell shape (see Fig. 4.1 below), whose peak is at a certain wavelength (i.e. color). At low temperatures, the peak corresponds to long waves, i.e. red light. Increasing the heat, the maximum of the curve moves to the right, where the short waves are, i.e. violet light, but up to certain temperature values the number of other colors is sufficient to ensure that the emitted light remains white. After these thresholds, the objects emit blue glow. If you look at the sky on a clear night, you will notice that the stars shine with slightly different colors: those tending to reddish are colder than the white ones, which in turn are colder than the blue ones. These gradations correspond to different stages of evolution in the life of the stars, as they consume their nuclear fuel. This simple identity card of light was the starting point of quantum theory, as we will see in more detail in a little while.

How fast does light travel?
The fact that light is an entity that "travels" in space, for example from a light bulb to our retina, is not entirely

intuitive. In the eyes of a child, light is something that shines, not that moves. But that is just the way it is. Galileo was one of the first to try to measure its speed, with the help of two assistants placed on top of two nearby hills who spent the night covering and discovering two lanterns at predetermined times. When they saw the other light, they had to communicate it aloud to an external observer (Galileo himself), who made his measurements by moving at various distances from the two sources. It is an excellent way to measure the speed of sound, according to the same principle that a certain amount of time elapses between seeing lightning and hearing thunder. The sound is not very fast; it goes at about 1,200 kilometers per hour (or 330 meters per second), so the effect is perceivable to the naked eye. For example, it takes 3 seconds before the thunderbolt comes from a lightning bolt that falls one kilometers away. But Galileo's simple experiment was not suitable to measure the speed of light, which is enormously higher.

In 1676, a Danish astronomer named Ole Römer, who at the time worked at the Paris Observatory, pointed his telescope towards the then known Jupiter satellites. These were called "Galileans" or "Medici" because they were discovered by Galileo less than a century earlier and dedicated by him to Cosimo de Medici. He focused on their eclipses and noticed a delay with which the moons disappeared and reappeared behind the big planet; this small-time interval depended mysteriously on the distance between Earth and Jupiter, which changes during the year. For example, Ganymede seemed to be early in December and late in July. Römer understood that the effect was due to the finite speed of light, according to a principle similar to that of the delay between thunder and lightning.

In 1685, the first reliable data on the distance between the two planets became available, which, combined with the precise observations of Römer, allowed him to calculate the speed of light. It resulted in an impressive value of 300,000 kilometers per second, immensely greater than that of sound. In 1850, Armand Fizeau and Jean Foucault, two skilled French experimenters in fierce competition with each other, were the first to calculate this speed with direct methods on Earth, without resorting to astronomical measurements. It was the beginning of a chase race between various scientists in search of the most precise value possible, which continues to this day. The most accredited today, which in physics is indicated by the letter c, is equal to 299792.458 kilometers per second. We observe incidentally that this c is the same that appears in the famous formula $E=mc^2$. We will find it several times, because it is one of the main pieces of the great puzzle called the universe.

Quantum Physics and the Law of Attraction

It is often hard to understand how the universe works; how you can get what you want, and how sometimes you just do not seem to get it. The Law of Attraction and Quantum Physics work together to create equilibrium in the universe. It is important to understand both of them, so that you can understand how the universe works.

The Law of Attraction – along with Quantum Physics – boils down to a very basic concept These are the facts that are important to remember when dealing with the Law of Attraction, so that you know how you can apply the law and what it means.

When you hear the phrase, 'Like Attracts,' it means exactly the way it sounds. The way you are, your attitude, your hopes and your dreams, are going to attract things similar to you. The type of energy you bring into the universe is the same kind of energy that attracts you.

Think of the moments when you were angry, upset, and running late. The more upset and frustrated you were during the day, the later you seemed to be running. The more you worried about being late, irritated or angry, the more you saw that you actually gave yourself more reasons to be upset, frustrated or late. Think now of a good day you have had in your life—a day when everything seemed to be going your way. You might be excited and happy, and there seemed to be nothing that could bring you down. The more you concentrated on these happy and excited emotions, the more you noticed, the more you were going to be happy and excited.

This is the fundamental idea behind the Law of Attraction—Like Attracts Like. The more you concentrate on good and positive things, the more the world gives you good and positive things.

This concept has been around for a long time, but it has only recently become popular, as more and more people begin to understand that the Law of Attraction is actually Quantum Mechanics, a theory of how the universe works. Quantum Physics teaches us that nothing is set, that there are no limits, that everything is vibrating energy. This Energy is under the control of our feelings. It is shaped, conformable and moldable. It is different from simply wishing and hoping, it boils down to believing. In order to make the Law of Attraction work for you, you must believe that the universe will send you the things you really want.

The Law of Attraction could end up being one of the simplest laws you have ever come across. When you fully understand it and are able to take advantage of it, you can find that you can have everything you have ever dreamed of.

The Law of Attraction is something that tells a person to draw things to themselves by concentrating on certain things. It has a relationship with Quantum Mechanics, which explains that there is nothing definite and there are no limits. According to Quantum Physics, all is made up of vibrating energy. The Law of Attraction and Quantum Physics are therefore both related and, in fact, interrelated.

According to Quantum Physics and the Law of Attraction, people are the creators of their own universe. The universe is made up of building blocks, not rigid like Newton's classical physics, but fluid and constantly changing, like quantum physics.

The Quantum Law of Attraction, therefore, is that because everything is always evolving and fluid—and, in reality, because the universe is made up of these dynamic and changing energies—everything can be attracted to any person, simply by concentrating. The likelihood that something will happen to someone is very high, as long as it is something they are focused on.

According to quantum physics, every person is part of the creation of the universe. That person focuses on issues and attracts them, and according to the issues they are concentrated on, these items are brought to each person. Therefore, the world is affected by our feelings. It is not something that is set—in stone—it is something that is movable and influenced by people's thoughts and by what they believe in.

For each person, this means that their dreams may become a reality. All they need to do is to focus on the things they want, and the things they have always wanted, and they are going to be able to draw opportunities to themselves much better than they might think they would do. In reality, bringing things to an individual is the only way to obey both quantum physics and the Law of Attraction at the same time. Focusing on the things you want and keeping them at the forefront of your mind is the best way to make sure you are motivated to do those things. You will find that you can do the things you believe in more easily. It is not always easy to believe that you can have whatever you want—but this is the foundation of the Law of Attraction.

According to the Law of Attraction, we attract everything that we constantly focus on. If we think about the relationship between the Law of Attraction and quantum physics, quantum physics explains that nothing in this world is fixed and there are no limitations. Quantum physics also states that all that exists in the universe is vibrating energy.

If you really want to fulfill your dreams and get out of the feeling of being trapped, you need to believe that everything in this universe is energy, and that this energy resides in a state of possibility. You have to allow the rule of attraction to be enforced in order to be successful. Remember, we are the builders of the universe. According to Newton's classical physics, the universe is made up of discreet building blocks. These blocks are solid, and they cannot be changed.

Quantum physics provides an explanation that there are no separate parts of the universe. Everything exists in fluid form and tends to change from time to time. Physics imagines this world as a deep ocean of energy that keeps coming into existence and disappearing from this universe.

People living in this world are changing the energy with their thoughts. It is therefore true that one can easily create what he or she wants to achieve. In short, human beings are primarily responsible for the achievement of their goals and the destruction of their desires.

The best thing to understand is that quantum physics has made us the creators of the universe. It is all energy around us.

You must have read Einstein's famous formula. The formula was introduced in 1905 and goes as follows:

$E = mc^2$

The above formula clearly explains the relation between energy and matter. Energy and matter can be modified quickly. In short, all that exists in this universe is energy, and energy is constantly evolving. Our thoughts have a great impact on this energy. Energy can easily be created, molded, and formed by our thoughts. We can easily turn the energy of what we think into the energy of what we really want to be.

Quantum physics is also known as the physics of possibility. This theory is contrary to the common idea that the outside world is real, and the inside world is fable. It says that whatever happens inside ultimately determines what happens outside. The world in which we live is created by our thoughts.

Nothing is fixed in this world, as mentioned earlier. Therefore, we need to realize that as we focus on our thoughts and on what we want to draw to ourselves, we can easily get what we want. Assume that "it can happen" and it will always happen.

The Law of Attraction and its strong connection to quantum physics will allow you to enjoy the success and achievement of your desires. Remember that good things will happen to people just because they believe they will.

The Law of Attraction and Quantum Physics are closely related. The Law of Attraction notes that through our thoughts and actions, we manifest reality. And not surprisingly, quantum mechanics will explain the Law of Attraction.

The most neglected and misunderstood branch of science at present is quantum physics. Quantum physics looks deeply into the structure of our existence and seeks to explain how the micro influences the macro, and how to grasp the origin of the Law of Attraction.

Although quantum physics is still not complete, due to the lack of resources to see deep enough to know anything, what has been discovered so far is adequate to understand the Law of Attraction in the world of thought.

One of the most significant discoveries in quantum physics is that matter can function as a particle or as a wave. Let me clarify that. A particle is a solid matter—it can only be in one position at a time, so you can still find its spot. However, a wave is not a finite point.

What quantum mechanics has now discovered, through observation, is that when very small particles—called electrons—are fired through two slits, they behave like particles. Each electron picked up a slit, went through it, and hit the back of the screen.

The result of firing hundreds or thousands of these, was a two-slit pattern. However, if the electrons were NOT detected when going through the slits, a broad interference pattern was formed on the back of the screen, which is the effect caused by the wave. In addition to this, the pattern showed interference from the slits, which further proved that the electrons passed through the slits as waves, not as solid particles.

So what does that mean for us? Our act of perception, feeling and emotion has an effect on the environment.

When scientists tried to track the electron to predict where it would go, they found that wherever the observer wanted it to end up, it was where it would show up. The consequences of this are equally enormous; our hopes, thoughts and beliefs literally shape the subatomic world around us!

Obviously, the power of our thoughts, emotions, desires, and values affect change, and constructing reality is just what the Law of Attraction tells us. Now that you have some scientific background, you might be able to put aside your current beliefs and give it a try. If by any chance you were told that you could have everything you wanted, would that at least be worth a try? Suspend your disbelief and be astounded.

Photoelectric Effect: Einstein's Theory

When electromagnetic radiation of the appropriate frequency is made to hit the surface of a metal such as sodium, the metal emits electrons. This phenomenon of emission of electrons from certain materials, which include several metals and semiconductors, by electromagnetic radiation is known as the photoelectric effect. This effect can be demonstrated and studied with the help of a set-up like the one shown in Figure 1.1.

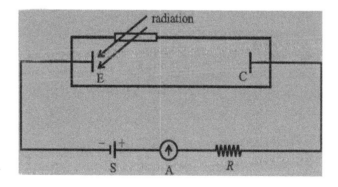

Figure 1.1: Set-up to study and analyze photoelectric effect; E is the emitting surface while C is the collecting electrode; A is a current-measuring device; S is a DC voltage source whose polarity can be reversed; R denotes a resistor; the actual circuit may not be as simple as shown here.

A metallic emitting electrode (E) and a collecting electrode (C) are enclosed in a vacuum chamber in which a window admits electromagnetic radiation of appropriate frequency to fall on E. A circuit made up of a source of EMF (S), a resistor (R), and a sensitive current meter (A) is established between E and C. The polarity of S can be changed so that C can be either at a higher or lower potential concerning E.

Features of Photoelectric Emission

This arrangement can be used to record several exciting features of the photoelectric emission. If for a given intensity of the incident radiation, the potential (V) from C to E is positive, then all the electrons emitted from E are collected by C, and A registers a current (I). This current remains almost constant as V is increased because all the photoelectrons are collected by C whenever V is flattening. This is known as the saturation current for the given intensity of the incident radiation.

This entire phenomenon of a current being recorded due to the emission of photoelectrons from E is dependent on the frequency (v) of the radiation. If the frequency is sufficiently low, then photoelectric emission does not occur, and no photo-current is recorded. For the time being, we accept that the frequency is high enough for the photoelectric emission to take place and we refer back to Figure 1.1. If holding the frequency and intensity of the radiation constant, one now reverses the polarity of S and records the photocurrent with increasing magnitude of V. One finds that the photo-current persists but decreases gradually until it becomes zero for a value V = −Vs or the potential of C with respect to E. The magnitude (Vs) of V for which the photocurrent becomes zero is called the stopping potential for the given frequency of the incident radiation. This is shown graphically in fig. 1.1.

The lower of the two curves shown in Figure 1.2 describes this variation of I with V for a given intensity (J1) of the incident radiation, and the frequency v being also held constant at a sufficiently high value.

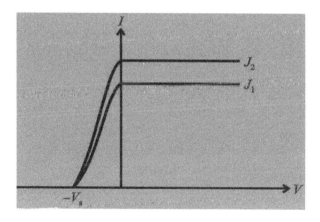

Figure 1.1: Graphical representation of the characteristics of photoelectric emission; the variation of photocurrent I with applied voltage V is shown for two values of radiation intensity, J1, and J2 (> J1), while the frequency v is held constant; the stopping potential Vs is independent of power.

If now the experiment is repeated for some other value, say J_2 for the radiation intensity, then a similar variation is obtained, as in the upper curve of fig. 1.1, but with a different value of the saturation current, the latter being higher for $J_2 > J_1$. However, the stopping potential does not depend on the intensity since, as seen in the figure, both curves give the same value of the stopping potential.

On the other hand, if the testing is repeated with different values of the frequency, keeping the intensity fixed, one finds that the stopping potential increases with frequency (Figure 1.2). One finds that if the frequency is made to decrease, the stopping potential is reduced to zero at some finite value (say, v_0) of the frequency. This value of the frequency (v_0) is found to be a characteristic of the emitting material and is referred to as the threshold frequency of the latter. In fact, no photoelectric emission from the material under consideration can take place unless the frequency of the incident radiation is higher than the threshold frequency. Moreover, for $v > v_0$, photoelectric emission does take place for arbitrarily small values of the intensity. The effect of lowering the intensity is simply to decrease the photo-current, without stopping the emission altogether.

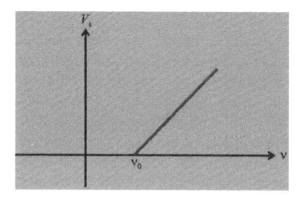

Figure 1.2: Variation of stopping potential with frequency. No photoelectric emission takes place if the frequency is less than the threshold value v_0, no matter how large the intensity may be.

The Role of Photons in Photoelectric Emission

All these observed features of photoelectric emission could not be accounted for by the classical theory. For instance, classical theory tells us that whatever the frequency, photoelectric emission should occur if the intensity of radiation is high enough, since for a high intensity of radiation, electrons within the emitting material should receive sufficient energy to come out, overcoming their binding force.

It was Einstein who first gave a complete account of the observed features of the photoelectric effect by invoking the idea of the photon as a quantum of energy, as introduced by Planck in connection with his derivation of the formula for the black body spectrum.

While the photons in the black body radiation were the energy quanta associated with the standing wave modes, similar considerations also apply to propagate radiation. Indeed, the components of electric and magnetic field intensities of propagating monochromatic electromagnetic radiation vary sinusoidally with time. Once again, a propagating mode of the field can be looked upon as a quantum mechanical harmonic oscillator of frequency, say, ν. The minimum value by which the energy of the radiation can increase or decrease is once again high, and this increase or decrease can once again be described as the appearance or disappearance of an energy quantum, or a photon, of frequency ν. Moreover, such a photon associated with a progressive wave mode, carries a momentum just like any other particle such as an electron. By contrast, a quantum of black body radiation energy has no net rate. The terminologies for energy and momentum of a photon of frequency ν are the de Broglie relations by now familiar to us:

$$E = h\nu, \quad p = \frac{h}{\lambda}, \quad (16\text{-}42)$$

where λ stands for the wavelength of the propagating monochromatic radiation and where only the magnitude of the momentum has been considered.

When monochromatic radiation of frequency ν is made to be incident on the surface of a metal or of a semiconductor, photons of the same frequency interact with the material, and some of these exchange energy with the electrons in it. This can be interpreted as collisions between photons and electrons, where the power of the photon engaged in a crash is transferred to the electron. This energy transfer may be sufficient to knock the electron out of the material, which is how photoelectric emission takes place.

Bound Systems and Binding Energy

A metal or a semiconductor is a crystalline material where a large number of atoms are arranged in a regular periodic structure. The electrons in such material are bound with the entire crystalline structure. In this context, it is essential to grasp the concept of a bounded system. For instance, a small piece of paper glued onto a board makes up a set system, and it takes some energy to tear the piece of paper off the board. If the power of the network made up of the paper separated from the board is taken as zero, since in the process of energy accounting, any energy can be given a pre-assigned value, since power is undetermined to the extent of an additive constant, and if the energy required to tear the paper apart is E, then the principle of conservation of energy tells us that the power of the bound system with the paper glued on to the board must be –E since the tearing energy E added to this initial energy gives the final power 0.

As another example of a bounded system, consider a hydrogen atom made up of an electron 'glued' to a proton by the attractive Coulomb force between the two.

Once again, it takes energy to knock the electron out of the atom, thereby producing an unbound electron separate from the proton. The power of the divided system, with both the proton and the electron at rest, is considered zero by convention, in which case the expression gives the energy of the bound hydrogen atom with the electron in the nth stationary state. Note that this energy is a negative quantity, which means that positive energy of equal magnitude is necessary to tear the electron away from the proton. This method of knocking an electron out of an atom is known as ionization. It can be accomplished with the help of a photon, which supplies the necessary energy to the electron, and the process is called photoionization.

In exactly a similar way, a hydrogen molecule is a bound system made up of two protons and two electrons. Looking at any one of these electrons, one can say that it is not bound to any one of the two protons but to the pair of protons together. In fact, the two electrons are shared by the pair of protons and form what is known as a covalent bond between the protons. Once again, it takes some energy to knock any one of these electrons out of the hydrogen molecule.

The minimum energy necessary to separate the components of a bound system is called its binding energy. Upon receiving this amount of energy, the components separate from each other, without acquiring any kinetic energy in the separate configuration. If the bound system receives an amount of energy greater than the binding energy, then the extra amount goes to generate kinetic energy in the components. In this context, an interesting result relates to the situation when one of the components turns out to be much lighter than the other. In this case, the extra energy is used up almost entirely as the kinetic energy of the lighter component.

Incidentally, when I speak of a bound system, I tacitly imply that it is to be looked at as a system made of two components. The same system may be looked at as one made up of more than two components as well. For instance, in the example of the piece of paper glued to the board, the components I have in mind are the paper and the board. But, given a sufficient supply of energy, the board can also be broken up into two or more pieces, and then one would have to think of a system made up of more than two components. In fact, the board and the piece of paper are made up of a large number of molecules, and the molecules can all be torn away from one another. Similarly, all of the two electrons and the two protons making up the hydrogen molecule can be pulled away from one another, for which a different amount of energy would be required as compared to the energy necessary to produce just one separate electron from an ion. We call the latter the binding energy of the electron in the hydrogen molecule.

The Theory of Relativity

It is impossible not to link the theory of relativity with Albert Einstein, its creator, perhaps the greatest scientific icon of the last century. Through the special version of his theory, Einstein succeeded in showing that the rules of electricity and magnetism do not depend on how the person who experiences it moves, contrary to what Newton's physics had concluded. He also showed that electromagnetic waves, unlike the other known waves at that time, exist even in a vacuum. Later on, his general relativity theory explained the "anomalous" changes observed in the orbit of Mercury. Furthermore, with no other tools than geometry and an axiom from the time of Galileo, Einstein correctly predicted that light suffers gravitational attraction, despite the fact that the particles that describe it do not have mass, and that light is bluer the closer it is to a body with mass.

These predictions that appeared in Einstein's original article were gradually confirmed more than half a century ago in different experiments and observations that catapulted Einstein to fame never before seen in the scientific media. For the changing society of the early twentieth century, the fact that using only mathematics incredible aspects of nature could be described, was a pleasant surprise. Perhaps not that it was an unprecedented experience, but few expected such a revolutionary change in a paradigm that had lasted for centuries. Before Einstein, the Newtonian description of gravity was revered for its simplicity and universal validity.

Relativity was soon discovered to hold more surprises. In 1915, the same year of its presentation in public, the theory revealed the possible existence of astrophysics bodies. Physics with stellar masses that do not let even light escape, or black holes. These matter-eaters beasts immediately became the most seductive and enigmatic celebrities of relativity for a wide audience. Soon after, Einstein showed that his theory predicted the existence of gravitational waves, deformations of the cosmic tissue that move at the speed of light, as if they were very fast space earthquakes. As if this were not enough, it was soon realized that a detailed scientific description of the history of the universe was possible based on the equations of relativity.

Despite the interest of these last entries, for several decades of the last century the study of black holes, gravitational waves and cosmology were seen as nothing more than curious theoretical works in more conservative scientific circles. The reason was and has been that it is difficult or impossible to make direct confirmations about these theories, and that it is difficult to make precise relativistic calculations to be able to compare them with indirect observations.

However, ingenious (and sometimes just lucky) experiments have accumulated vast indirect evidence.

For example, in cosmology, the discovery of cosmic background radiation, predicted and described by the Big Bang cosmological model as remnant radiation from an early time when the universe was very hot, was an important building block in the consolidation of cosmology. This discovery was award-winning. Despite this, there were still doubts on whether the big bang model, based on general relativity, was correct to describe this radiation. Fortunately, a decade later it was confirmed that radiation is not uniform throughout the cosmos and that the measurement of these inhomogeneities matches what the theory had predicted. Celebrated with the Nobel Prize in 2006, this discovery did not leave many options open. The cosmology described by relativity is the most appropriate description. A more detailed measurement of the expansion of the universe soon led to the last greatest cosmological discovery: the universe grows faster and faster (Nobel Prize 2011). Additionally, almost everything can be perfectly adjusted from general relativity.

The confirmation of the existence of black holes and gravitational waves does not have a shorter history. For one thing, black holes have always represented a singular scientific nuisance. The fact that the theory indicates that the gravitational force at the heart of black holes is infinite indicates a serious problem: right there the theory of general relativity is no longer valid. Thus, for some time it was considered that black holes were a trick that mathematicians played on us.

However, theorists argued that, like stars made entirely of neutrons, black holes were corpses of stars larger than ours. The confirmation of the existence of neutron stars in the 1960s and indirect observations of the movement of stars around dark regions led to the certainty that there are many black holes in the universe, and that they can have masses millions of times that of the Sun.

Additionally, black holes, while not the typical cosmic vacuum cleaners that we paint, do absorb large amounts of matter from their cosmic neighborhood, creating around them a ring of incandescent radiation emitting material called an accretion disk. This radiation predicted by the theory has been confirmed especially in the center of galaxies like ours. In the Milky Way, the movement of a group of stars around the center of the galaxy, "chased" by astronomers since the 1990s, has revealed that there lives a relatively small and dark body with a mass of almost 3 million times that of the Sun and that it emits radiation according to predictions for a hole. As if this were not enough, in 2011 an accretion disk was observed with emission of X radiation, consistent with the predictions of black holes with masses of billions of times that of our star, absorbing material from a quasar.

The last series of indirect observations of black holes has to do with the gravitational waves. Although these can occur with any accelerated movement, even a clap, according to the theory only gravitational waves generated by violent cosmic events are capable of producing gravitational waves that we can detect on Earth, with sensors so sensitive as to measure deformations of the space of a thousandth the size of a proton or less. Gravitational wave detection led to the award of 2017 Nobel Prize. Supercomputers with sophisticated programs managed to show that the signals, according to relativity, were only consistent with the collision and mixing of two black holes with masses equivalent to a few dozen times that of the Sun. All at once, two of the most controversial predictions of relativity were proven! And the evidence continues to accumulate today for the purpose of exploring possible deviations from the predictions of Einstein's theory and finding astronomical applications of the study of gravitational waves.

Not only gravitational waves and black holes are under the scrutiny of research. As we said, the cosmological model of the big bang can explain all observations of the dynamics of the universe under very simple assumptions on the geometry of space-time and an assumption on the content of the cosmos based on recent measurements, which is: 5% is matter like that of our planet, 27% is a type of matter called dark matter that does not emit light, and 68% of cosmic content is a form of energy nicknamed dark energy that causes the accelerated expansion that we observe.

The biggest mystery is that no one has the slightest idea what dark matter and dark energy are. They are nothing that we have been able so far to observe directly, although there is sufficient indirect evidence to affirm that there are such "substances" or something that has the same effects. There are those, however, who are convinced that we must slightly modify the basic equations of general relativity to understand the true nature of these dark entities. Others consider that only the immensely challenging search for the compatibility of general relativity and quantum mechanics will help us dissipate our doubts.

This mix brings us back to black holes. The inside of the black holes, being not observable, is totally unknown. All we know is that gravity must be so strong inside that it could have effects on the smallest particles, comparable to the effects of quantum forces that, according to particle physics, govern their behavior. If so, it is possible that a form of quantum gravity is manifested there, which we must theorize based on what we have verified in the last century.

Even for those least interested in the fundamentals of gravity, the theory of relativity today offers important modern tools. In addition to being relevant for the global positioning system (GPS), it is crucial in astronomical observations. The deflection of light due to its passage near galactic formations, stars, planets, etc. causes an apparent shift in the position of the stars and galaxies with respect to the real one. It is not the only effect. If there is a galaxy behind a very massive astrophysical body, the deflection of light beams emitted by the galaxy in all directions can be deflected towards us around the contour of the galaxy.

Astrophysical "nuisance." This effect is called gravitational lensing. Gravitational lenses not only allow us to characterize what is behind the observable objects that cause the deflection of light, but also, when they occur in regions where there are no visible obstacles, they show us properties of unobservable objects, such as black holes and dark matter formations, which have not yet been fully described.

Relativity, despite its age, remains a developing treasure, whose questions and responses pose current challenges that are likely to become the basis for the future discoveries and paradigm shifts like the one Einstein witnessed.

Quantum Physics and Waves

These are two usual aspects of Einstein's field practice and Maxwell's electric field. One more approach to take a look at the quantization cycle is to at first overhaul field conditions corresponding to mathematical administrators who consolidate those numerical standards.

There are a large number of establishments for the use of quantum field speculation. In any case, a typical hypothesis of traditional style convictions, which is one of our best (non-esteem) things for the contemplation of nature. Second, quantum field theory can speak to (perceptions, adjusted suspicions) the creation and crumbling of particles, non-logical proportions of a quantum material. Third, the quantum field theory is relativistic inherently, and "mysteriously" handles complex issues that plague even the quantized atom hypotheses.

Be that as it may, no, quantum fields are not viable with any kind of effect. Quantum fields are significant. In quantum field speculation, what we see as particles is basically a fascinating field of the quantum field.

Quantum electromagnetism is an unmistakable "convenient" theory of the quantum field. There are two fields in it: the electromagnetic field and the electron field. Along these lines, for instance, what we normally observe as an electron-focused electron is a sure association in quantum electrodynamics between an electromagnetic field and an electron field, where the electromagnetic field loses quantum incitement, and the electron field assimilates its strength and performance power.

How would you explain the idea of the quantum wave of the problem?

What is a Wave?

- It generally quantifies the range of a sine wave, i.e., the distinction between the broken-down waveforms.

- Frequency measures how often the sine wave is reset in a second.

- Quantity quantifies the size of the frequency scale over zero levels.

- The stage decides the situation of the point on the wave in the second situation in the space, in recurrence units.

Wavelength Measurement

- Size A

- Wavelength λ

- Category Shift $\Delta\varphi$

Interruption

It is extremely useful to utilize wave impedance to discover superfluous allotments. If two wave surfaces are suspended, their non-wave pinnacles may abbreviate (gainful impedance) while confronting a higher value, and the body, by and large, will radiate a wave. The example of ruinous and ensuing impedance in space makes it simple to envision recurrence.

It is an element of material science that the lifting power is not identified with the dense concentrations of particles and the fragile set of a single atom in a machine at some random time. The capacity of the wave, despite everything, demonstrates a true quantum object. This is one motivation behind why, sometimes, it implies that 'all cells are isolated.'

Quantum theory can recognize the likelihood of a specific result. Which of these expectations is finally expected in the essential inclusion of the overall political race and the regions? Just a couple of appraisals under similar conditions show cautious dissemination of chances, which is additionally taken out from Schrödinger's framework.

$i\hbar\, \partial/\partial t\, \psi(r, t) = (-\hbar 2/2m\, \Delta + V\,(r,t)\psi(r,t)$

All designs to date have demonstrated the following: square modulus | ψ | 2 of the state work ψ refers to the probability of getting a quantum object during special position t and all the various boundaries contained in ψ.

Slender Film Disruption

The obvious impacts of impedance are not restricted to the twofold edged computations used by Thomas Young. The impact of a small block of film is caused by the light that shows the two zones isolated by a range equivalent to their size.

The "film" in a space can be water, air, or some other indistinguishable or strong fluid. In splendid light, the obvious impedance impacts are restricted to films with the extent of a couple of microns. The notable model is an air pocket cleaner film. The light reflected from the air pocket is a two-wave lift, one obvious on the front surface and the other on the back. Two waves show spread and interruption into space. The size of the cleaning film decides if these two waves can meddle with help or in a dangerous manner. The full test shows that considering only the recurrence λ, there is a useful film thickness impedance equal to $\lambda/4$, $3\lambda/4$, $5\lambda/4$, and destructive interference for thickness $3\lambda/2$.

As the white light enlightens the cleaning film, the hidden gatherings are viewed as different frequencies moving through destructive obstacles and isolated from the show. The reflected light is consistently shown as a comparing shade of the irradiated recurrence (e.g., when a red light is produced with a damaging impedance, the splendid light shows up as cyan). Thin-oil films produce a near impact on water. In Nature, the quills of winged creatures, including peacocks and fowls, just as the shells of specific creepy crawlies mirror light as the shade of the significant changes with a modified state.

This is achieved by restricting intelligent light waves from misleadingly planned structures or by the wide variety of show posts. Subsequently, abalone pearls and shells sparkle from the restriction brought by presentations from different pieces of mother-of-pearl. Stones, such as opal, show the flickering impacts of gleaming that originate from dissipating light from the commonplace instances of round particles.

There are numerous uses of hindering light impact mechanical impacts. Inclusion's main enemy is the camera's center focal point that focuses on little estimated films, and recovery records taken to make the impedance of a risky showcase of clear light. Constant progressed inclusion, which incorporates different slight film layers, is made to transmit light only within a narrow range of wavelengths and thus fill as recurrence channels. Multilayer textures are also used to enhance the mirror presence on infinite telescopes and laser optical gaps. Genuine interferometry methodology measures small changes in related isolation by observing turning shifts in light-hindering plans. For instance, the state of Earth in the obvious parts is reflected in the optical waves segments using interferometry methods.

Wave-fields and Interference

Like in nature, it is conceivable that understanding has developed dynamically and we think that for us people it is over. However, we are not a gathering with a decent comical inclination. But this can dramatically decrease: it is restricted to living creatures with neurons. Few can guarantee that cell life alone is conceivable, or that plants have it (and I likewise incorporate reptiles).

Regardless, we are powerless as far as where we are gaining little ground, which is 1.5 billion after the primary acknowledgment of Great Climacteric. Neurophysiology, with all the neuroscience involved, will discover nothing like that.

Other than that, there is a favorable position in our stockpile that we, regardless of all, have not yet used: the information loupe. At its most honed point, a wonderful spot in the region was found a few million years back by the Great Climacteric sign. We will see where evolution switched gears in its neuronal endeavors and began another cycle.

Cognizance Polychrome

Since the mid-1960s, tasks have been progressing at NASA and at different institutions for the proof of intelligent life in space - SETI exercises. If these endeavors pay off, we may receive an answer to our messages. One such message will tell us there is life on a farthest planet, and where it is.

The Heisenberg's Uncertainty Law

The uncertainty standard, otherwise called the Heisenberg Principle of Uncertainty or the Principle of Indeterminacy, was created by the German thinker Werner Heisenberg (1927). It states, in principle, that both the shape and the speed cannot be known at the same time.

Conventional experience does not understand this principle. The condition of a vehicle and its position can be known at the same time.. The complete law specifies that the result of exposure is equivalent to, or is more noteworthy in position and speed than the base or persistent worth ($h/(4\mu)$) in which Pl Plk constants or roughly $6.6/10^{-34}$ seconds). The impact of shakiness just applies to little fields of iotas and sub-nuclear particles.

Any attempt to precisely calculate the speed of a molecule under a particle, for example an electron, could affect it in such a way that it will not be possible to calculate its position at the same time. These discoveries or inventions have nothing to do with the insufficiency of estimation, cycle, or surveying devices, in light of close common contact of particles and waves with subatomic size.

Every molecule has a wave; all particles have a wave-like nature. The molecule is most regularly discovered when waves are enormous or substantial. Also, the more noticeable the frequency, the more prominent it becomes, and the molecule pressure sets. Only a spotless wave has an infinite length; despite the fact that its reference molecule has a specific position, it has a particular speed.

A molecule can be anywhere, despite the fact that it has a specific speed. A precise estimation of a lone difference alludes to the overall vulnerability, while figuring different factors.

The idea of vulnerability is likewise communicated as far as elements and molecule particles. Molecule pressure is equivalent to the result of its size and speed. This way, the vulnerability's impact is comparable to or greater than $h/(4\alpha)$ on the strength and position of the molecule. The guideline applies to other equal sets of perceptions, for example quality and time. The vulnerability result has yet to be determined from the intensity and exposure over time of the most prominent or equivalent examination $h/(4\beta)$. Due to a precarious particle, there is a close association between the uncertainty of the radiation measurement and the danger of the unstable frame, as it prompts stable advancement.

Quantum Super-Positioning

Any time you play the guitar and hear the notes, you feel the impact of the waves. The melodies of every arrangement consolidates as they reach your ear. When you toss a pebble on a lake, something very similar occurs: there are waves that meet on the shoreline.

Sound waves and water waves rise up, completed by singular wave focuses that produce another wave.

These two conditions have one shared factor: the rotating waves consolidate, resulting in a point-by-point proportion that delivers another wave.

Waves can represent molecules, electrons, and a few different occupants of the quantum universe. Their mobile pinnacles and valleys may have total qualities estimated by quantum resources.

For instance, the electrons circling a particle do not exist anywhere known to man as it does the Earth when it orbits the sun. Preferably, they are set in an orbital haze of chance. This space cloud is a practical 3D quantum wave, comprising mountains and valleys that change over time and speak of the opportunity to have electrons in a given space.

The calculation of this wave shifts according to the quality of the electron. A surface can be made where two quantum waves - speaking to two degrees of electron vitality - are assembled, prompting another example of pinnacles and valleys. This changes where the electron is well on its way to be found and the particle's remarkable structures can be influenced.

Steve Rolston, President of the Department of Environment at the University of Maryland, clarifies why our everyday experience does not have a quantum scale.

It is not unexpected to say that maybe an electron has two unique energies simultaneously, or that it is in a few places simultaneously in this kind of amplification. In the case that you consider an electron as only a molecule, this will not be clear. In any event, when you think about an electron as an all-encompassing element, the overlay is more straightforward. Waves - including wave super-positions - are found in numerous spots simultaneously.

The setting may from time to time appear to be oddly extraordinary. For example, one might put an apple close to an orange and calling it a banana. However, it is valid.

At a point, electrons act like particles. Consider small balls, with an example of two arrangements of locators with a set behind each space. At that point, the identifier follows the unsettling influence, as though each electron ventured like a wave between the two limits.

A quantum object moves as a wave and a molecule. Slides are focused on singular particles, yet the following example is that of a wave.

Quantum Computing

Classical computers, the thoughtful ones we use every day, use memory made up of bits. Bits speak either ones or zeros; on or off. Everything computers do, from messing around to sending an email, originates from controlling those ones and zeros.

A quantum PC is another kind of PC that uses the irregular properties of quantum material science to solve problems that are unthinkable for standard computers. They do this by utilizing quantum bits (qubits) rather than bits. Like bits, qubits can speak ones or zeros. What makes them extraordinary is that a qubit can be a one, a zero, or a superposition of both. That implies that a qubit can be both one and zero simultaneously, making quantum computers exponentially more dominant than their ordinary partners.

By utilizing superposition, quantum computers can solve problems that would be unthinkable or take a considerable number of years to solve. Quantum computers drastically outflank old-style machines in calculations, including enormous quantities of similarly potential arrangements.

Because of their quality at dissecting mixes, quantum computers will likely be applying to break codes and to streamline complex frameworks. Researchers likewise expect that quantum computers will have the option to precisely display solutions at a subatomic scale, providing a useful asset to science and to material science research.

Superposition is incredible, secretive, and fragile. The most significant boundary to the construction of working quantum computers is that qubits must be in a super-cooled, disconnected state, or they decode themselves and lose their quantum "enchantment."

Quantum computers are sitting on the edge of common sense.

What is the harsh deal with quantum computing? Imagine you are in a huge workplace and you need to recover a portfolio left on a work area picked aimlessly in one of several workplaces. You would need to stroll through the structure, opening entryways each in turn to discover the folder. A regular computer would need to clear its path through long strings of 1's and 0's until it landed at the appropriate response.

Its underlying foundations can be followed back to 1981 when Richard Feynman noticed that physicists consistently appear to run into computational problems when they attempt to recreate a framework in which quantum mechanics would happen. The computations, including the conduct of molecules, electrons, or photons, require an enormous amount of time on modern computers. In 1985 in Oxford, England, the first depiction of how a quantum computer may function surfaced with David Deutsch's hypotheses. The new gadget would not exclusively have the option to outperform the modern machines in speed, but it could also perform some legitimate tasks that conventional ones could not.

This research started investigating the development of a gadget, and with the thumbs up and extra subsidizing of AT&T Bell Laboratories in Murray Hill, New Jersey, another individual from the group was included. Subside Shor discovered that quantum calculation could extraordinarily speed up considering whole numbers. It is more than just a stage in a small-scale computing innovation, and it could offer bits of knowledge into certifiable applications, for example, cryptography.

In our computers, circuit sheets plan with the goal that a one or a zero is spoken by varying measures of power. The result of one plausibility has no impact on the other. In any case, an issue arises when quantum speculations are presented, the results originate from a solitary bit of equipment existing in two separate substances, and these realities cover each other influencing the two effects on the double. These problems can probably get the best quality out of the new computer; however, it is conceivable to program the results this way along these lines, that adverse impacts counterbalance themselves while the positive ones strengthen one another.

This quantum framework must have the option to program the condition into it, confirm that is the calculation, and concentrate the results. A succession of heartbeats could be used to show the particles into a usable example in our arrangement of conditions.

Another plausibility from MIT's Seth Lloyd is the proposal to use natural metallic polymers (one-dimensional particles made of rehashing iotas). The vitality conditions of a given molecule would control by its connection with neighboring iotas in the chain. Laser heartbeats could be used to send flags down the polymer chain, and the two closures would make two individual states of vitality.

A third proposition was to replace natural atoms with gems wherein data would be put away in the precious stones in specific frequencies. The nuclear cores, turning in both of two states (clockwise or counterclockwise), could be modified with a tip of a nuclear magnifying lens, either "perusing" its surface or changing it, which obviously would be "expressing" some portion of data stockpiling. "Monotonous movements of the tip, you could, in the long run, work out any ideal rational circuit, " DiVincenzo said.

This force includes some significant pitfalls, in the sense that these states would need to remain isolated from everything, including a wanderer photon. These outside impacts would gather, making the stray framework track, and it could even pivot and wind up moving in reverse, causing regular mistakes. New hypotheses have arisen to conquer this.

One path is to keep the calculations moderately short to lessen odds of blunder. Another is to reestablish repetitive duplicates of the information on discrete machines and take the standard (method) of the appropriate responses.

It would, without a doubt, surrender any preferences to the quantum computer. Thus AT&T Bell Laboratories have created a mistake amendment strategy in which the quantum bit of information would be encoding in one of nine quantum bits. It would be the ensured position that the quantum state would enter before being transmitted. The health of the molecule could be resolved basically by watching the far edge of the particle, since each side contains the definite inverse extremity.

Researchers today tend to center the entryways that would transmit the data around this single quantum rationale door and its plan of segments to play out a specific activity. One such entryway could control the change from a 1 to a 0 and back, while another could take two bits and make the outcome 0 if both are the equivalent and one is unique.

These entryways would be lines of particles held in an attractive snare or single molecules going through microwave holes. This single door could be developing in a year or two, yet a legitimate computer must have a great many entryways to get down to earth. Tycho Sleator of NYU and Harald Wein Furter of UIA take a look at the quantum rationale entryways as straightforward strides towards making a quantum rationale arrangement.

These systems would nevertheless be columns of entryways collaborating. Laser bars sparkling on particles cause progress starting with one quantum state, then onto the next, which can adjust the sort of aggregate movement conceivable in the exhibit. Thus, a particular frequency of light could be used to control the associations between the particles. One name given to these clusters is "quantum-dab exhibits," where the individual electrons would be kept to the quantum-dab structures, encoding data to perform scientific activities from straightforward expansion to the figuring of those whole numbers.

The "quantum-speck" structures would be based upon propels, taking the shape of little semiconductor boxes, whose dividers keep the electrons restricted to the small district of material, another approach to control how data is preparing. Craig Lent, the primary researcher of the undertaking, bases this on a unit consisting of five quantum dabs, one in the inside and four. At the parts of the bargains, electrons would be burrowing between any of the two locales.

Hanging these together would make the rationale circuits that the new quantum computer would require. The distance would be adequate to make "paired wires" made of columns of these units, flipping the state toward one side, creating a chain response flip every one of the groups states down along the wire, much like the present dominoes transmit latency.

How Quantum Physics Affects You

Quantum material science is ostensibly the best scholarly victory throughout the entire existence of human development. Yet, to the vast majority it appears as though it is excessively distant and theoretical. This is, to a great extent, a self-dispensed injury concerning physicists and pop-science authors. When we talk about quantum material science, we typically stress the unusual and nonsensical marvels: Schrödinger's cat in a superposition of being both "alive" and "dead," Einstein's issue with God playing dice, the peculiar significant distant connections of the quantum trap. These things are stimulating since they are extraordinary, yet examining them in the lab requires disengaging fundamental quantum frameworks. It is not easy to perceive any association between these wonders and the existence of regular people.

However, quantum material science is surrounding us. The universe, as we probably are aware, runs on quantum rules. Although the old-style material science and the quantum physical science are different, there are bunches of recognizable, ordinary wonders that owe their reality to quantum impacts. Here are a couple of instances of things you likely run into in your everyday daily existence without realizing that they are quantum:

Toaster ovens:

As you toast bread or a bagel, the red gleam of a warming component is an exceptionally recognizable sight for the vast majority of us. It is additionally where quantum material science got its beginning: explaining why hot articles shine that specific shade of red is the issue that quantum physical science was developed to solve.

The shade of light produced by a hot article is a case of such a straightforward, all-inclusive marvel that is catnip for hypothetical physicists. Regardless of what an item is made of, if it can endure being warmed to a given temperature, the range of light it transmits is equivalent to some other substance. Such widespread conduct attracted a ton of genuinely brilliant physicists in the last part of the 1800s. However, none could solve the issue.

You count up all the shades of light that an article may radiate and give every one of them an equivalent portion of the warm energy contained in the item. The issue with this is there are significantly more approaches to emanate high-recurrence light than low-recurrence light, which recommends that as opposed to a beautiful warm res gleam, your toaster oven ought to shower x-beams and gamma beams everywhere in the kitchen.

The answer to this issue was found by Max Planck, who presented the "quantum speculation" (giving the inevitable physics its name) that light must be discharged in discrete lumps of energy. For high-recurrence light, this energy quantum is more significant than the portion of warmth energy allocated to that recurrence, and consequently, no light is radiated at that recurrence. This cuts off the high-recurrence light and prompts an equation that coordinates the watched range of light from hot items to incredible accuracy.

Thus, every time you toast bread, you are taking a look at where quantum material science began.

Glaring Lights:

Old-school brilliant lights make light by getting a bit of wire sufficiently hot to transmit a splendid white shine, making them quantum similar to the toaster oven. On the off chance that you have bright light bulbs around, either the long cylinders or the more up to date twisty CFL bulbs, you are getting light from another progressive quantum measure.

Back in the mid-1800s, physicists saw that each component in the occasional table has a novel range: if you get a fume of molecules hot, they radiate light at a small number of discrete frequencies, an alternate example for each component. These "ghastly lines" were immediately used to distinguish the creation of obscure materials, and even to find the presence of already mysterious elements. Helium, for instance, was first identified as a formerly unknown ghostly line in light from the Sun.

While this was powerful, no one could clarify it until 1913 when Niels Bohr got on Planck's quantum thought (which Einstein reached in 1905) and presented the primary quantum model. Bohr proposed that there are individual unique states wherein an electron can joyfully circle the core of a particle, and that molecules assimilate and discharge light as they move between those states. The recurrence of the light incorporated or produced relies upon the energy contrast, in the manner presented by Planck, giving many discrete frequencies for a specific iota.

This was an extreme thought, yet it worked splendidly to clarify the range of light transmitted by hydrogen, and the x-beams produced by a broad scope of components, and quantum material science was hightailing it. While the cutting-edge image of what happens inside a molecule is different than Bohr's underlying model, the central thought is the equivalent: electrons move between the great states inside iotas by engrossing and emanating light of specific frequencies.

This is the central thought behind fluorescent lighting: Inside a bright light bulb (either long cylinder or CFL), there is a tad of mercury fume energized into a plasma. Mercury happens to discharge light at frequencies that generally fall in the apparent range in a way that can trick our eyes. The light looks white. Take a look at a bright light bulb. You will see a couple of detailed shaded pictures of the bulb, where a brilliant bulb gives a persistent rainbow smear.

Along these lines, any time you utilize bright lights to light your home or office, you have quantum material science to thank for it.

Physics:

While Bohr's quantum model was valuable, it did not first accompany a physical explanation concerning why there ought to be individual states for electrons inside molecules. That was not explained for almost ten years, and ended up being the reason for the most groundbreaking mechanical upset of the century.

The extreme thought that gave Bohr's exceptional energy states a physical premise originated from Louis de Broglie, a French Ph.D. understudy from a distinguished family. He recommended that as Planck and Einstein had presented a molecule like nature for light waves (where a light emission can be thought of as a flood of "light quanta" each conveying one unit of energy for that recurrence), there may be a relating wave-like conduct for particles like electrons. If you give electrons a frequency, you find that there are "standing wave" circles where the electron wave finishes a whole number of motions as it circumvents the nucleus. These have precisely the correct energies to be Bohr's exceptional states in hydrogen.

This wave conduct can be legitimately estimated, and it immediately was so in both the US and the UK. Those waves drove Erwin Schrödinger to his wave condition, and hence to one of the fundamental ways to deal with the full present-day physics of quantum mechanics.

The wave idea of electrons significantly changes our comprehension of how they travel through materials, prompting our advanced understanding of energy groups and band holes inside materials. We can utilize this material science to control semiconductors' electrical properties. By putting together pieces of silicon with the specific right mix of different components, we can make little semiconductors that structure the essential elements used to handle advanced data.

In this way, every time you turn on your physics, you are misusing the wave idea of electrons and the unprecedented control of materials that permits. It may not be the hot sort of quantum physics, yet every cutting-edge physics needs quantum material science to work appropriately.

Natural Compass

There are instances of quantum material science in regular day to day existence. If you feel that only humanity has been fortunate enough to use quantum physics, you are wrong! As per speculations by researchers, flying creatures like the European robin use quantum physics to move. A light-delicate protein called cryptochrome contains electrons. Light photons, after entering the feathered creature's eyes, hit cryptochrome, and extremists are delivered. These extremists empower the fledgling to "see" an attractive guide. Another physics proposes that the mouths of the winged creatures contain gorgeous minerals. Shellfish, reptiles, creepy crawlies, and even a few warm-blooded animals utilize this sort of natural compass. You may be astounded to know that the kind of cryptochrome used by flies in their routes has likewise been found in the natural eye! Its utilization is hazy.

Semiconductor

Semiconductors have far-reaching uses as they intensify or switch electrical signs and electrical force. Taking a look at semiconductors' structure, we would understand that a semiconductor comprises layers of silicon related to different components. Many of these make microchips, and these CPUs structure the force to be reckoned with the apparent multitude of mechanical devices that have gotten vital to human presence. Had quantum physics not become an integral factor, these chips would not have been made, and neither would work areas, tablets, physics, or cell phones..

Laser

The guideline on which laser works depends on quantum physics. The working of lasers includes unconstrained outflow, warm emanation, and fluorescence. An electron, when energized, will bounce to a high-energy level. It will not remain in the high-energy level for quite a while; it will hop back to the lower energy state that is steadier, and, subsequently, it will produce light. Outer photons likewise influence the iota's quantum mechanical condition at a recurrence related to nuclear progress.

Microscopy

Electron microscopy has improved with the hidden standards of quantum physics, which, in combination with electronic microscopy, has improved the imaging of natural examples. In differential impedance contrast microscopy, an example of obstruction is made by the light emission, which is then dissected. Across the board, with quantum physics, microscopy has greatly improved, and, in this manner, a lot of data can be obtained.

Global Positioning System (GPS)

Exploring obscure areas has never been simpler as it has been with the guide of quantum physics. When using a cell phone for the route, the GPS on your telephone gets signals from various satellites.

The separation and time between your location and your destination are determined by computing the signals from various satellites. Each satellite has a nuclear clock, which depends on quantum physics to work.

Attractive Relativity Imaging

Attractive Relativity Imaging, otherwise called Nuclear Magnetic Relativity (MRI), includes the inversion of the electrons' twists in hydrogen cores. Thus, fundamentally, we are discussing shifts in energies, which is only one of the uses of quantum physics. The investigation of delicate tissues can undoubtedly be completed with the utilization of MRI. Due to quantum physics, some dangerous diseases can be treated.

Media transmission

Correspondence has been made very simple due to the significant part of quantum physics. Fiber optic media transmission has made possible two-way and speedy correspondence. The fiber optic media transmission is possible because of lasers, which are gadgets of quantum physics.

Super Precise Clocks

Dependable timekeeping is more important than your morning alarm. Tickers synchronize our mechanical world, keeping things like securities exchanges and GPS frameworks in line. Standard timekeepers utilize the ordinary motions of physical items like pendulums or quartz gems to deliver their 'ticks' and 'tocks.' Today, the most exact checks on the planet, nuclear tickers, can utilize quantum physics standards to quantify time. They screen the particular radiation recurrence expected to take electrons to leap between energy levels. The quantum-rationale clock at the U.S. Public Institute of Standards and Technology (NIST) in Colorado just loses or increases a second every 3.7 billion years. Furthermore, the NIST strontium clock, uncovered recently, will be that precise for 5 billion years—longer than the current age of the Earth. Such super-touchy nuclear tickers help with GPS routes and broadcast communications.

The exactness of nuclear tickers depends mostly on the number of particles used. Kept in a vacuum chamber, every particle autonomously gauges time and contrasts itself with its neighbors. If researchers pack multiple times more iotas into a nuclear clock, it becomes various times more exact—however, there is a breaking point on the number of molecules you can crush in. Scientists' next giant objective is to use ensnarement to upgrade accuracy effectively. Caught molecules would not be distracted with nearby contrasts and would rather exclusively gauge time's progression, viably uniting them as a solitary pendulum. That implies that including multiple times more molecules into an entrapped clock would make it numerous times more exact. Ensnared timekeepers could even be connected to frame an overall organization that would quantify autonomous time of an area.

Uncrackable Codes

Customary cryptography works using keys: A sender uses one key to encode data, and a beneficiary uses another to unravel the message. Nonetheless, it is hard to eliminate the danger of gossip, and access can be undermined. This can be fixed utilizing possibly strong quantum critical dispersion (QKD). In QKD, data about the key is sent through photons that have been haphazardly captivated. This limits the photon, so it vibrates in just one plane—for instance, here and there, or left to right. The beneficiary can utilize enraptured channels to interpret the key and afterward use a picked calculation to safely encode the message. The mystery information gets sent over typical correspondence channels, yet nobody can decipher the message except the specific quantum key. That is precarious because quantum directly decides that "perusing" the spellbound photons will consistently change their states. Any attempt at listening in will make the communicators aware of a security breach.

Today, organizations, such as BBN Technologies, Toshiba, and ID Quantum, use QKD to plan super-assured organizations. In 2007 Switzerland evaluated an ID Quantum item to give a sealed democratic framework during a political race. What is more, the principal bank move utilizing ensnared QKD started in Austria in 2004. This framework vows to be exceptionally secure, since, in such a case that the photons are snared, any progressions to their quantum states made by gatecrashers would be promptly obvious to anybody checking the key-bearing particles. Yet, this framework does not work over enormous separations. Up until now, entrapped photons have been sent over a most significant break of around 88 miles.

Super-Powerful Computers

A standard physics encodes data as a line of parallel digits, or pieces. Quantum physics supercharge handling power since they use quantum bits, or qubits, which exist in a superposition of states. Until they are estimated, qubits can simultaneously be both 1 and 0.

This field is still being developed. However, there have been positive developments. In 2011, D-Wave Systems uncovered the D-Wave One, a 128-qubit processor, followed a year later by the 512-qubit D-Wave Two. The organization says these are the world's first economically accessible quantum physics. In any case, this case has been met with doubt, since it is as yet indistinct whether D-Wave's qubits are trapped. Studies delivered in May discovered proof of ensnarement yet just in a little subset of the physics' qubits. There is additionally vulnerability about whether the chips show any dependable quantum speedup. NASA and Google have collaborated to frame the Quantum Artificial Intelligence Lab dependent on a D-Wave Two. Furthermore, researchers at the University of Bristol a year ago snared one of their conventional quantum chips to the Internet so anybody with an internet browser can learn quantum coding.

Quantum Microscope

In February, a group of analysts at Japan's Hokkaido University built up the world's first entrapment improved magnifying instrument, utilizing a method known as differential impedance contrast microscopy.

This kind of magnifying tool fires two light emissions at a substance. They measure the reflected bars' impedance design—that change whether they hit a level or uneven surface. Using entrapped photons significantly improves the data the magnifying lens is registering.

The Hokkaido group figured out how to picture an engraved "Q" that stood only 17 nanometers high with exceptional sharpness. Comparative strategies could improve space science instruments called interferometers, which superimpose various rushes of light to break down their properties more readily. Interferometers are used in the chase for extrasolar planets, to test stars, and to look for the swells in space time called gravitational waves.

The European robin might be a quantum average.

The attractive field is encompassing the fowl impacts on how long these cryptochrome extremists last. Cells in the bird's retina are believed to be touchy to the snared extremists' presence, permitting them to adequately 'see' an attractive guide dependent on the atoms.

However, this cycle is not fully seen, and there is another alternative: birds' attractive affectability could be because of little gems of beautiful minerals in their noses. All things considered, if ensnarement is truly at play, tests propose that the sensitive state should last any longer in a 10,000 foot than in even the best frameworks. The attractive compass could likewise be material to specific reptiles, shellfish, creepy crawlies, and even a few well-evolving creatures. For example, a type of cryptochrome utilized for the route in flies has likewise been found in the natural eye, even though it is muddled if it is currently or previously valuable for a comparative reason.

Conclusion

The microscopic world has its own rules, which, as David Wheeler wrote, sound impossible. Some think that there must be a more reasonable and realistic understanding of the reality behind the quantum theory. One of the viewpoints for the advancement of quantum physics itself is the understanding of many universes. Wheeler says that you never know this for sure, until the science confirms or refutes. He says that the universe comprises not only the everyday reality but also the rest of the world, about which we learn more as science develops.

Quantum physics is not the first stage and may not be the last step in the continual development of our knowledge of the universe. For the time being, it is the most progressive view of the reality of humanity. It is not just about the micro-world, but also about our daily facts. Despite this reasoning, Newtonian mechanics appears to be a reliable method for other practical applications. Quantum mechanics will accompany modern science. "Would this process be endless? Would our knowledge ever be complete? But these are questions from a different field, the field of science philosophy."

When electrons are intertwined, it means that the measurement shows the opposite of their spin signs. This interconnection occurs when the particles form in a single process. According to the exclusion principle of Pauli, every quantum system has different components. Any electron may turn out to have a positive or negative spin, but the signs are usually the opposite. The spin of one electron can be enough to automatically determine the other's spin, as in the two-slit experiment.

According to quantum theory, the second electron's spin sign is definite and opposite to the first. When one electron is measured, both electrons' wave function collapses regardless of the distance between them. The electrons demonstrate their final interconnection as part of what is known as a single quantum system. After that, electrons are no longer connected, and in the future, they will be able to acquire all properties independently. They claim that they can get entangled with new particles with which they later interact, including photons. It can continue with some delay, depending on the distance, as time and speed can never be measured with absolute accuracy. Quantum theory proposed that simultaneous interconnections would take place at any distance. Einstein denied the notion of jamming, but modern experiments showed otherwise. There are no methods for measuring time and speed with absolute accuracy, but the instruments' efficiency is improved. Scientists believe that the rate of contact approaches the speed of interconnections. This interaction seems to be infinite velocity, i.e., both particles at the same time acquiring exact features regardless of distance (non-locality). "It is obvious why Einstein so slowly dismissed the Entanglement theory."

Nancy Patterson

Made in the USA
Monee, IL
02 April 2023